IDEAS AND THE MATTER

**WHAT WILL WE BE MADE OF
AND WHAT WILL
THE WORLD BE MADE OF?**

MARINELLA FERRARA
GIULIO CEPPI

DIPARTIMENTO DI DESIGN

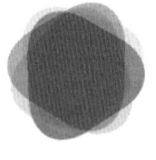

Material Design Culture Research Centre is the research center and network dedicated to materials for design of the Design Department of Politecnico di Milano. It deals with the relation design-materials among design history and actuality. The center carries researches on the Italian Culture of Materials Design, intended as the capacity of the Italian design to interpret materials and technologies in order to generate products and environments innovation. The research team is composed by: Marinella Ferrara (coordinator), Giampiero Bosoni, Giulio Ceppi, Sebastiano Ercoli, Chiara Lecce, and Andrea Ratti.

www.madec.polimi.it

INTRODUCTION

MARINELLA FERRARA
GIULIO CEPPI *pg.7*

PART I.

THE IDEAS AND THE MATTER

Extension

THE CONFINES OF THE BODY, THE ENDS OF THE PROJECT
FRANCESCO SAMORÈ *pg.16*

HUMAN IN THE MOLECULAR AGE
GIUSEPPE TESTA *pg.18*

Osmosis

BALANCE BETWEEN THE INTERIOR AND THE EXTERIOR
MASSIMO FACCHINETTI *pg.35*

THE THIRD PARADISE
MICHELANGELO PISTOLETTO *pg.37*

MATER MATERIA
CLINO TRINI CASTELLI *pg.44*

ATOMIC DESIGN AND THE ARTIFACTUAL ELEGANCE
ROBERTO CINGOLANI *pg.64*

THE GROWING LAB. FUNGAL FUTURES
MAURIZIO MONTALTI *pg.71*

Inclusion

THE ADVANCE FROM THE EXTERIOR TOWARDS THE INTERIOR
FORTUNATO D'AMICO *pg.91*

RESEARCHING THE FUTURE
STEFANO MARZANO *pg.94*

CITY-WORLD AND WORLD-CITY
MARC AUGÈ *pg.105*

PART II.

CRITICAL THINKING FOR A NEW FRAMEWORK

NEW VISIONS FOR A DIFFERENT DESIGN ANTHROPOMETRY
GIULIO CEPPI | pg.109

TOWARD THE BIOCENTRIC ERA. OBSERVING DESIGN HYBRIDIZATION
CHIARA LECCE | pg.132

ETHICS AND DESIGN IN THE 21ST CENTURY. NOTES
GIAMPIERO BOSONI | pg.149

PART III.

OPEN UP TO MATERIAL INNOVATION

FUNDAMENTALS OF MATERIAL DESIGN CULTURE
MARINELLA FERRARA | pg.157

SHIFTING TO "DESIGN-DRIVEN MATERIAL"
MARINELLA FERRARA | pg.173

DESIGN-DRIVEN MATERIAL INNOVATION METHODOLOGY
MARINELLA FERRARA
CHIARA LECCE | pg.186

AUTHORS' BIOGRAPHIES | pg.206

INTRODUCTION

MARINELLA FERRARA
GIULIO CEPPI

This book aims to present the founding research undertaken by the Politecnico di Milano's Material Design Culture Research Centre (Madec). Founded in 2014, Madec obtained the Design Department's support during its first year by being granted the *Fondo di Ateneo per la Ricerca di Base* (FARB 2013) for "Fundamental/Foundational/ Exploratory Researches that are strategically assessed for scientific growth in a research department".

Madec's research is at the heart of the *Material Design Culture* concept, defined at the beginning of the research as "the whole that includes knowledge, beliefs, ethics, habits and any other skills acquired in the relationship between design and materials" (Ferrara & Lecce, 2015a). Material Design Culture is based on design expertise in the adoption, interpretation and invention of material technical advances, in order to develop technological and scientific opportunities for productive activities, and to make human experiences with objects and living spaces the most meaningful, rich and rewarding. The Material Design Culture is very rich in Italy, thanks to the craftmen knowledge, design skills and industries's capabilities on manifacturing, and also thank to the debate on design qualities.

Madec's work team is composed of: Giampiero Bosoni, full professor of Interior Design and design historian who, since his cooperation with Vittorio Gregotti and his work for *Rassegna* magazine in the 1980s, became the spokesman for the issue on the relationship between design and materials; Giulio Ceppi, senior researcher of Industrial Design and an architect who, in the 1990s, cooperated on the materials identity issue with Ezio Manzini and Antonio Petrillo in the Domus Academy Research Centre; Andrea Ratti, associate professor of Industrial Design, designer in the nautical sector and expert on composite materials based on fibres; Chiara Lecce, PhD in Interior Architecture and Exhibition Design, and Sebastiano Ercoli, PhD in Industrial design. The team is coordinated by Marinella Ferrara, associate professor of Industrial Design, focused on the relationship between design and technology for innovation, authors of a number of publications that link micro-stories to macrostructures, and for rethinking the relationship between design and materials as a dynamic of the socio-technical innovation process.

Starting from the key concepts of Material Design Culture, and considering the Italian cultural context within which Madec was conceived and operates, the founding research was organized in two phases.

The first phase focused on the identification of the *Italian Materials Design Culture* through a historical retrospective study. Starting from the past and studying Italian design's consolidated practices has been a useful strategy to deeply understand the specific Italian approach "to interpret materials generating innovation", in both linguistic and deployment terms, and even through its evolution in the international context. This historical study, by means of prominent scholars' original research contributions on the topic, was published in the *AIS/Design. Storie e Ricerche*, the scientific journal of the Association of Italian Design Historians. This issue entitled "Italian Material Design: learning from history", edited by Giampiero Bosoni and Marinella Ferrara (2014), is rich in contributions, discoveries and insights on Italian Material Design Culture. It is largely based on the original archive documents that report the history of small and large companies, as well as episodes of handcrafted excellences from different Italian regions during the XX century. The collection of contributions proves the definition of an Italian way founded on the historical dialogue between technique and aesthetic and on the particular attention of Italian designers for designing and communicating materials' meaning and values.

Madec's second research phase, to which this book is the *compendium*, focused on the future design perspective of Material Design Culture's evolution on the basis of the emerging perception of the world which is facing an increasingly complex scenario of scientific advances, socio-economic problems and challenges.

In October 2014, in order to identify the actual evolution of the relationship between science knowledge and design, MADEC started a wide debate with a series of open lectures and seminars entitled "The ideas and the matter: What will we be made of and what will the world be made of?" (Ferrara & Lecce, 2015b; Lecce, 2015). This initiative was organized in partnership with the Giannino Bassetti Foundation[1] which is socially engaged on responsible innovation, and with the Cittadellarte-Fondazione Pistoletto[2] that promotes art's interaction with all the areas of human and social activity. Therefore, the initiative was conducted in parallel with the Final Synthesis Design Studio of the Interior Design Degree (academic year 2014/2015) coordinated by Giulio Ceppi, with Fortunato D'Amico, Massimo Facchinetti and Francesco Samorè, acting as discussant facilitators in the seminars. Many other colleagues from different Politecnico di Milano departments have been participating in the lectures as discussants, and thus enhancing public debate.

The integration between research and didactic activity has been very stimulating. The students were invited to work in the Final Session Workshop to solve project tasks, thus defining new areas of research and reflection, including ideas from the lectures. The Workshop's objective was, therefore, to re-interpret the concept of interior design in the context of the great social and technological transformations in progress and the expansion of the design scales that confront the designer in the face of new and complex responsibilities. It was stimulating to see the students' project results evidencing the influence of the different concepts assimilated from the lectures. All the heterogeneous contributions stemming from professional experiences, theories developed and applied research in several disciplines (medicine, biotechnology, physics, engineering, anthropology, art, architecture and design) which feed design concepts, have had the merit of significantly broadening and deepening the themes. The debate demonstrated that design is a powerful tool for mediation in very different specializations, and an innovation tool addressing the challenge from very different perspectives. Design plays a role of providing new meanings in a world full of opportunities, yet also with problems to be solved. Envisioning design for emerging markets is crucial in order to create or enter the market.

Narrating and considering advanced scenarios in design schools is a challenging task that facilitates the advancement of the expertise domain needed to create meaningful disruptive ideas. But today, this is no longer sufficient to respond to innovation's complex dynamics. There is a lack of knowledge on how to apply innovation in a complex system as a cyclical and repeatable process. Small and medium-sized enterprises are looking for new markets and new applications but are limited to new business partners or new processes. The innovation process is long, and such companies may not move on to expand the application base. They need the support of technology centres and technology transfer companies to facilitate this on a wide scale. Material experts from the design world can also play an important role in facilitating the required research network. Madec initiated the very first steps in this direction by organizing other activities: the design workshops "Material Design. Regenerated plastics for food packaging" (2014), "Material Design. Contaminazioni ceramiche" (2015), and "Marble Visions" (2016), all in partnership with enterprises in order to understand what today companies actually need in the materials and product sector to bring about innovation and let it flow proposing new product/service concepts; and, the networking workshop "Stone reinforced ecoconcrete. Open innovation Project" (2016) to stimulate a new research approach for material sector starting from new visions of material, production cycle and product concept; all this together with an open access web site project (www.madec.polimi.it) to publish updated news on interesting case studies and research that is focused on materials design in the international scenario. These activities yielded suggestions for advancing a *material thinking* approach and an *open innovation* discourse.

A new opportunity for the Madec research came from the European tender "Capabilities for design-driven innovation in European SMEs" (2015) founded by EASME (European Agency for Small and Medium-Sized Enterprises). During this project, set up by the consortium formed by D'Appollonia[3], MIP[4] and ADI[5], Madec was involved in drawing up a module on the relationship between material and design and proposed the "Design for Materials" module. Designing this module, based on the reasoning and reflections inspired by previous activities, has allowed MADEC's research to be finalized into an easy tool that contributes to the dissemination of a set of information and skills to accelerate the material innovation processes among design and production professionals, companies and intermediate organizations. The Design for Materials module includes information on the several factors to be considered for developing or choosing a material for consumer products. It proposes the concept of material performances, a concept that takes into account sensory perception, consumers' experiences and cultural values, as well as functionality related to the materials' properties. It also enabled the proposal of the Design-driven Material Innovation Methodology as a tool of open innovation, to manage a design process where different actors, such as researchers, suppliers, creative communities and customers are becoming deeply engaged.

1.
http://www.fondazionebassetti.org/

2.
http://www.cittadellarte.it/en

3.
http://www.dappolonia.it/en

4.
http://www.mip.polimi.it/en

5.
http://www.adi-design.org/

The Design for Materials module, integrated with design, design management and process engineering modules, makes up the "Design for Enterprises" - a three year programme of training courses to support design-driven innovation. These courses involve the use of old and new materials, new production tools and processes and approaches to better meet customers' needs. The programme provides for the implementation of 50 courses (short and full-term) in 29 different European and associated countries from 2016 to 2018.

In conclusion, after describing the Madec research process, this book shall publish selected contributions from the public debate it has been developing, essays on the theoretical reflection that was derived from the public debate within the research group, and phenomological and both theoretical research on design approaches in order to develop the Design-driven Material Innovation Methodology, all grouped into three different chapters.

In the first chapter, the book collects selected trans-disciplinary contributions that emerged from the series of seminars "Ideas and the matter: What will we be made of and what will the world be made of?" (held from October to December 2014). Following the structure of the seminars, contributions are divided into three sections: Extension, Osmosis and Inclusion. These words/concepts act as metaphors of the complex, and often conflicting, relationship between polarities such as interior-exterior, body-space and artifice-nature, that are increasingly characterizing the contemporary design world. We pioneered and coordinated the seminars based on these thematic areas. Each section is introduced by a discussant facilitator and publishes the text of the lectures.

Francesco Samorè introduces the section *Extension* or from the inside of our bodies to the outside, starting from how technologies and fields of knowledge such as genomics, bio technologies and nanotechnologies are changing the concept of life and body, but also on how it is possible to visualize the invisible and infinitely small to understand and modify it, rather than knowing and measuring everything that happens in our body with biometrics or the quantified self. The infinitely small redefines us and redefines itself, creating new grounds for design and social value.

The concept of Extension is addressed by the molecular biologist Giuseppe Testa, director of the Laboratory of Stem Cell Epigenetics of the European Institute of Oncology and European School of Molecular Medicine (Milan). Testa explains how our relationship with technology is changing today. The power of biotechnology deconstructs the body into its parts making visible what was once obscured, and blurs the borders between natural and cultural and internal and external perpetually, reshaping and redesigning the images of the human body and the ways in which this body is conceived.

Massimo Facchinetti presents the section *Osmosis* or the balance between the interior and the exterior, between our physical and energetic boundaries: skin as a psychic limit and place of narrative, from tattoo to intradermal patches. Just as all of today's architecture plays on the skin's theme: sophisticated technical systems and materials work on the bioclimatic and plant-based membranes and on the exchange between the internal and the external. It is the body theme of smart skin on the one hand, but also the delicate relationship between urban services and functions, public and private osmosis, between open and closed, between ephemeral and enduring. The concept of Osmosis, as generically defined by the dictionary is a gradual, often unconscious,

process of assimilation or absorption. This definition well fits the aim of this second section of lectures that try to give a correlation between different disciplines similar to an osmotic movement. The Osmosis concept was tackled by the artist Michelangelo Pistoletto, who explains his project *Third Paradise* concerning the conflict between *nature* and *artifice*. Another contribution is *Mater Materia* by the designer Clino Trini Castelli who, along with his work on color and material surfaces, addressed the role of a number of other "subjective" aspects of space, including light, sound, temperature, texture, and scent. In 1972, he coined the term *design primario* for this work, following Roberto Cingolani, a physicist and Scientific Director of the Italian Technology Institute (IIT) in Genoa. His work includes different fields of scientific research such as: *Advanced Robotics* or *Nanophysics*. Collaborating with the designer Chris Bangle, he explored the future of consumer applications for robotics and the approach to human-machine interaction that attempts to renegotiate the idea of "robot" away from that of industry and into an arrangement of emotional and functional interpretation. The last Osmosis essay is dedicated to the work of Maurizio Montalti, a young designer and founder of the Officina Corpuscoli studio in Amsterdam, an interesting example of design driven by materials experimentation. In his specific case, the material used is the fungus *Mycelium* which is able to grow on a wide variety of substrates with the potential of converting waste products into novel compounds, characterized by diverse qualities.
Fortunato D'Amico presents the section *Inclusion* i.e. the advancement of the outside inwards, of the environment loaded with information that is immersed within us and which increasingly takes on push logic: wearable technologies that change our relationship with the outer space, putting on (when not Inside) things and systems that were "out" before: consider goggle glasses, which seems to be like applying a layer and augmenting reality. As regards the city's macro and physical scale, it is also the agriculture that returns to the city, so the gardens become roofs, rather than the productive periphery that returns to the historical centres.
Our inner vision amplifies our capacity to perceive and absorb the outer. In the educational field, the encounter of different knowledge, before separated, could become an ideal procedure to implement participatory approaches and sharing processes. Designer responsibilities to society and the environment have to be constantly in evidence, especially today that we are facing a continuous growth of dumps also because of the uncontrolled production of design objects. Inclusion is faceted by Stefano Marzano and Marc Augé's tireless efforts to analyze the contemporary world with the tools of their relative professions and cultural approaches. Stefano Marzano brings his long lasting experience as Chief Design Officer and CEO of Philips Design in Philips between 1991-2011. Marc Augé, the famous French anthropologist, who opposes the "City World" of global business, tourists and architects for the "World City", the megacity where all differences become apparent - social, ethnic, cultural and economic.
In the second chapter, the book collects the interpretative picture of Madec members. Following a debate with a strong international and interdisciplinary character, researchers opened different insights on changes in practices with a wider vision of the "matter and materials" and "design and meta-design" concepts. The team was urged to reflect on the following points to enlarge the fundamentals of our knowledge, with a common idea of cross-pollination between disciplines:

- The depth of our look inside matter is augmented, the real matter on which we focus our look has changed and the places and actors of knowledge production have changed as well;
- The complex relation between technology and nature, considered hostile, could be skillfully managed by humans trough all the "meanings" that creativity could contribute to conceive and express the avoidance of obsolete models;
- Today, nature could become the measure and method for designing the artificial, guaranteeing sustainability and beauty, and therefore competitiveness;
- The knowledge gap that high specialization contributed to creating could be overtaken through a wider multidisciplinarity.

Open and shared knowledge is the only way that will enable us to propel ourselves toward the future. Thus, Giampiero Bosoni proposes a new interpretation of the wide relationship between ethics and design. Giulio Ceppi proposes a reflection on the future perspective of design, according to the extent to which the boundaries of human knowledge have been expanded from the macro-world to the micro-world. Chiara Lecce analyses the contemporary design scenario on the basis of a growing hybridization between natural science, engineering and design disciplines, thanks to a greater capacity of researchers to integrate and coagulate experiences under the umbrella of a more sustainable future.

In the third chapter, the scientific discourse focuses on the changing approach of product materiality and the evolution of research methods. Today, changes are opening new options for design action, new ideas and the definition of design approaches, contributing to the development of new methodologies. In this chapter, Marinella Ferrara traces the fundamentals of Material Design in the Italian design history, and analyses the changes in design discipline to arrive at the definition of *design-driven materials*. In the new dimension of the contemporary Material Design Culture and, together with Chiara Lecce, presents the *Design-driven Material Innovation Methodology* which provides for the integration of tailor-made materials during the design process, in order to create new scenarios of material and product concepts. In this methodology, design plays a role of giving new meanings through the design of materials and products with a critical approach. This is a mission that designers cannot forgo, following the recent successes of Design Thinking, which is opening up to the social innovation challenge and achieving creative solutions beyond the reach of conventional structures and methods.

In the research process, due to the understanding and interpretation of social needs, design is pushing users' desires towards the opening of new visions on use and the application of old and new materials systems, and delivers them to customers through new product characteristics and storytelling approaches. This innovation is able to give new meanings, new experiences and is opening new markets. On the subject of materials, this means considering markets and final users as the main drivers, focusing on new product concepts and on new applications that are able to generate unexplored design solutions as well as innovative behaviour in society and, in so doing, generating business.

Madec has initiated the very first steps to drive material design research towards an open innovation approach, but much more has to be done in order to fulfil this goal.

ACKNOWLEDGEMENT

Like any design process, also a research and a book are greatly influenced by the relationships that occur during its doing. We would like to express our sincere gratitude to all those who gave up their time to contribute to the research activities like meetings, seminars, workshops, and the series of lectures we have been organize. We are sorry that we have not been able to publish all the contributions. We are grateful to: Pietro Cecini from Roadrunnerfoot, Matteo Lai from Empatica, Chris Bangle from Chris Bangle Associates, Raymundo Sesma from Campo Expandito, Marco Baudino from Future Power and Tiziana Monterisi from Nova Civitas. Their contributions stimulated us to widen our research from various perspectives.

Thanks to Enrico Benco and Cristina Talon of GS4C for their inspiration, energy and valuable insights. Many concepts took shape during the preparation of the workshop networking "Stone reinforced ecoconcrete. Open innovation Project" organized in partnership during our research.

We are grateful to our colleagues for participating as discussants to the seminars the colleagues. Those are: Mario Bisson, Luisa Collina, Fiammetta Costa, Barbara Del Curto, Luca Guerrini, Maurizio Masi, Marina Parente, Silvia Piardi, Giovanna Piccinno, Dina Riccò, Valentina Rognoli, Paolo Volonté.

Our sincere thanks also goes to the Bionike and Moleskine for its sponsorship.

PART I.

IDEAS AND THE MATTER

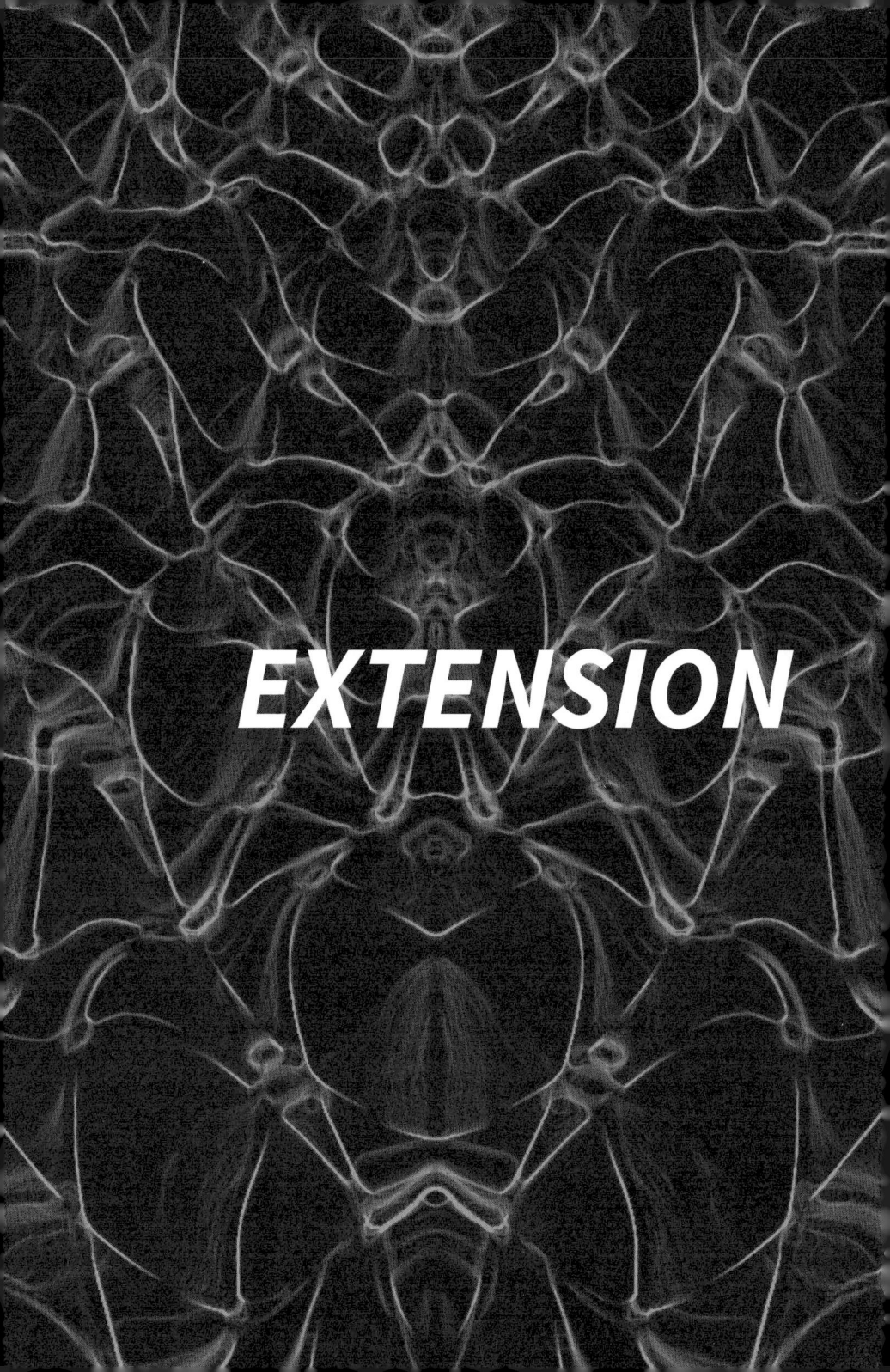

THE CONFINES OF THE BODY, THE ENDS OF THE PROJECT

FRANCESCO SAMORÈ

Fondazione Giannino Bassetti, Milan, Italy

The trajectories of innovation have so strengthened the adaptability of our look (its depth, width) as to put into discussion the way in which the designer "reads" – and consequently designs – the relationship between interior and exterior.
From as early as the 1920s, John Dewey had already sensed how technological societies would generate moments that resist the solution attempts in the ambit of the existing institutions. Without exceeding in the provocation, we can reasonably affirm that design is one of the "institutions" set up to spur said awareness in our time; and which naturally, as a discipline, is not safe from the effects that it contributes to illuminate.
As metaphor "transports" the meaning of words, so does design allow meanings between ambits of life that are in appearance distant and contributes to making the connections between them tangible.
This is why the *Estensione* chapter develops like an itinerary:
 - it was inspired by the *interior* of the human body, allowing students to learn from a prestigious geneticist how the molecular glance facilitates the design of life through genetics and studies its relationship with the *exterior* thanks to epigenetics;
 - it is followed taking its place at the confines of the body, reflecting on the ways in which technology, needs and desires unite in the design of innovative prosthetics
 - it has projected itself onto the environment, completing the extension by studying the link between intrinsic spaces of technology – transformed by sensors, domotics, robotics, the Internet of things – and the person.
And in fact, we are basically talking about relationships. The itinerary from the interior to the exterior is, as often occurs during the travel experience, circular. We may return to our departure point, but we now look at it throughout different eyes.
Here we come back to the "look", which should be considered in its daily nature: how genomics revolutionised the digital representation that allows us to see – and therefore manipulate – the bricks of the living being (DNA, epigenomes, cells) so that *wearables technologies* transform what is intimate into digital data, bringing it to the exterior: they measure heart beats, they inform us of our emotional reactions, they quantify our eating habits. Designing artificial objects means therefore, more than ever, designing the relationship between the interior and the exterior, between people and context.

It is a task to carry out in places in which knowledge is transmitted, because innovation is a surplus of knowledge that meets a surplus of power. It can change history and the mark it leaves is never neutral. A discerning question, awareness, knowledge of purposes, not distinct nor subordinate to the methods.

For the Fondazione Giannino Bassetti, which has for fifteen years promoted responsibility in innovation, meeting the Politecnico di Milano – with its lecturers, starting with Giulio Ceppi, and the students – has confirmed how the designer is becoming less and less a "supplier of means" and increasingly a "co-decider" regarding the purposes (Bassetti, 2004).

Reference

Bassetti, P. (2004). Nuova scienza e nuova politica. In Colombo, U. & Lanzavecchia, G. (Eds), *Scienza e tecnologia al di là dello specchio*. Milan: Libri Scheiwiller.

HUMAN IN THE MOLECULAR AGE

GIUSEPPE TESTA
IEO & Università degli studi di Milano, Italy

I think there are many interesting hints in the various keywords and titles making up this research: "Ideas and the matter", "Inside and outside", "What we will be made of", "What the world will be made of".
Let us begin therefore by asking ourselves what we will be made of and what we are partly already made of, with a brief introduction about myself. I work at the European Institute of Oncology that boasts one of the most important basic research components in Italy. My lab, in particular, is devoted to the study of the role played by genes and epigenomes. In the course of this conversation, we will learn how to confer a more precise meaning on these terms.
For the time being, I will only hint at the fact that we study the regulation of our genome within the context of brain development and two pathological contexts, on the one hand tumour diseases and on the other hand brain development diseases, particularly autism and intellectual disability.
At the same time, together with a colleague, we have founded, already many years ago, an academic program that studies science and biomedicine specifically as phenomena in themselves, with the aim to train a series of scholars with a dual skill, capable of working in a lab while simultaneously developing a fully-fledged competence in philosophy and sociology of science, in bioethics and in biojurisprudence. This is, therefore, the other side of my scientific activity, where I essentially deal with the manner in which biomedical and biotechnological innovation penetrate society and society, in turn, penetrates biotechnological innovation making it possible and expressing it in various ways.
Let us start therefore with a movie quotation to introduce the notion of "trailer" that has been the image I have associated with the need to move from inside to outside. What happens when we shoot a movie, at least certain types of movies, is this continuous camera movement, which allows precisely this probing of various scenic perspectives. Still sticking somehow to the movie context, I divided the essay into a prologue and three "acts", quoting to some extent the movies of the Dogma 95 movement, inasmuch as a characteristic of these movies was to grant viewers the impression of live action and the immediacy of events through a special use of the camera. I want to introduce this metaphor at once as a trace, stylistic as well, precisely because the three developments I will speak about have to do with bringing genomes, epigenomes and cells

from the inside to the outside, being developments currently underway in the here and now of our contemporaneity.
I accordingly want to give of these developments an absolute live impression, aesthetically as well.

PROLOGUE
The prologue, like any prologue worthy of its name, starts from afar and starts from Michel Foucault, an intellectual, an archaeologist of knowledge.
Among his several contributions, Michel Foucault writes what is possibly one of the most beautiful essays called *The Order of Things* (1966), which we retrace here as we speak ideally of *ideas and matter*. Within this context, we are going to speak of body, specifically human body, of matter, of disease.
The Order of Things is one of Foucault's most important essays in which he undertakes an archaeology of medical knowledge, i.e. he asks himself how we came to know medical things in different historical epochs. The following quotation is taken from the beginning of the book setting out the story of a doctor, called Pomme, who had treated a woman affected by what was then termed *hysteria*, suggesting for ten consecutive days, on a daily basis, ten hours of immersion in water. Pomme described the outcome of this approach thus:
"Membranous parts like pieces of wet parchment detached themselves with light pains and exited daily with the urine, whereas the right hand side ureter undressed in turn and came out in one piece through the same duct, the intestines at another time divested themselves of their inner tunic we saw coming out of the rectum, the oesophagus, trachea and tongue were in turn undressed, while other pieces were expelled either through vomit or through expectoration" (Foucault 1998, p.3).
Foucault juxtaposes, and this is in my view one of the greatest intuitions of this treatise, this type of description onto another description which is of course actually later, though less than one century so, where another doctor was observing the brain, again belonging to a patient affected by mental illness, and describes it instead thus:
"(...) their outer surface applied onto the small arachnoid sheet of the hard mother cleaves to this sheet now loosely, and then separates them easily, now tightly and intimately, in which case it is extremely difficult to detach them, their inner surface is only contiguous to the arachnoid with which it contracts no union, the false membranes are often transparent, especially when they are very thin, but usually have a whitish, greyish, reddish colour, and more rarely a yellowish, brownish and blackish colour, this matter often displays different nuances depending on the parts of the same membrane, the thickness of such accidental productions varies considerably, they are sometimes so faint that they might be compared to a spider's web." (Foucault 1998, pp3-4)
I do not think there is any need to point out that they are two methods, quite diverse, of describing bodily matter, a pathological matter in this instance.
Here, Michel Foucault raises the essential question serving as point of departure of the treatise. In other words, he asks and invites us to ask how drastically our way of looking at things has changed: "(...) who can ensure us that an eighteenth century doctor, the first one, did not see what he saw, but that a few dozen years have been enough for

the fantastic figures to disperse away and the space thus freed let the clear-cut slant of things reach up to the eyes ?" (Foucault 1998, pp.3-4).

Foucault, therefore, contrasts the first description, assimilated to a world peopled by phantoms, of these nearly ghostly figures of parts of matter proceeding outwardly with the secondo ne that shows at every phrase "the clear-cut slant of things". He intuitively realizes that this "the clear-cut slant of things" has to do with space, especially with what he terms the spatialisation of disease and body. In the process, he states something we might find surprising today, though in actual fact a thorough historical analysis highlights with utter certainty, namely:

"The space of configuration of the disease and the space of localisation of the bodily illness have only been superimposed in medical experience for a short period, the one that coincides with nineteenth century medicine and with the privileges conferred on anatomical pathology. An epoch that marks the sovereignty of the glance, since within the same perceptive field, following the same continuities and the same fault lines, experience reads in one shot the visible elisions of the organism and the coherence of pathological forms.

Evil expresses itself the first time precisely on the body, and its logical distribution is done straightaway for anatomical masses." (Foucault 1998, p.16)

Foucault intends to say that, actually, what seems nowadays a given, in the way we look at bodies, more specifically at ill bodies, is in reality no more than a very recent way of looking and thinking, one that historically coincides with the rise and disciplinary-institutional success of anatomical pathology, i.e. this spatialisation of symptoms and diseases on specific groups of things, specific groups of organs, specific groups of bodily forms.

Prior to that, there was what historians term classifying medicine, in which, "before being taken into the thickness of the body, medicine precisely received an organisation hierarchically divided into families, kinds and species."(Foucault 1998, p.16) There was in other words a whole list of diseases, relationships between diseases, connections between diseases. The disease was "essentially" perceived "within a space of projection without depth and coincidence without development."(Foucault 1998, p.18) A bit like the way the Earth was conceived before it was discovered that it was spherical. It is no coincidence that Foucault will say it is a *tableau*, i.e. a table gathering these representations and classifications of diseases that used to be thought of as natural and ideal species alike. There is a disease that is A, that is B, that is C, and these are natural species, as natural as the species of the world, so much so that genealogical trees were drawn that sought to re-apply to the discovery of diseases and to the narration of diseases the same type of thinking that aimed at organizing the things of the world, the matter of the world, the plants, animals, ourselves, etc.

It was a case, in this approach, "of natural and ideal species together, natural because diseases reveal their essential truths, and ideal to the extent that they never give themselves in the experience without alteration or disorder." (Foucault 1998, p.20).

But who brings this disorder and this alteration? It is brought by the body of the individual, it is when individuality breaks in, instead, that we will see a central trace of the bio-technological contemporaneity. That is why I wanted to introduce this prologue. In fact, Foucault goes on to say, "to this pure nosological essence – i.e. to this pure, quite abstract definition of disease – that un-procedurally defines and consumes its place within the order of the species, the sick person, here we name him for the first time, i.e. the individual body, adds as equal number of perturbations his dispositions, his age, his lifestyle, a whole series of events that compared to the essential nucleus manifests as accidents." (Foucault 1998, p.20)

"Those who describe a disease – it is one of the doctors of the time who writes that –, must see to the task of distinguishing the symptoms that necessarily accompany it and are peculiar to it from the accidental or fortuitous ones, such as the ones that depend on the sick person's temperament or age" (Sydenham, cited in Foucault 1998, p.20). Here Foucault concludes the chapter stating that "[...] paradoxically, the patient is, compared to what he suffers from, but an external factor, medical examination must not take him into account other than to place him in brackets." (Foucault 1998, p.20). This was, therefore, classifying medicine, this flat world of ideal entities of sicknesses, in which, when the patient, i.e. the individual in his specific body, was studied, it was necessary to track down those ghostly forms we have seen at the beginning, in an attempt to pursue their ideality and abstraction, removing what were the confusing data. What were they? The person's age, his sex, his temperament, words we no longer use, but which clearly had to be put somehow in brackets.

Keep in mind this prologue, as we will come back to it at different times, and I hope you will notice how different is at present our way of thinking, and yet how important it is, precisely because our way of thinking is different, for us to refer back to these precursors.

Think only of a word I think all of you at least hears, namely, personalized medicine, i.e. the emphasis to personalize any type of treatment or preventative scheme that does exactly the opposite of what classifying medicine used to do. Far from placing the individual in brackets, in fact, he is placed, firstly, at the centre of an epistemological project that hopefully will later turn into therapeutic project.

ACT I

Having concluded the prologue, we may now begin with the first act; more specifically, we will start from the genomes. As usual, we could have chosen several points of departure, but I thought that the DNA somehow was an already partly familiar element. In 2008, British parliament was debating a rather controversial law that concerned the possibility of creating "new things", i.e. embryos in which parts of human cells and animal cells were mixed together, in particular the nucleus of a human cell inserted in the cytoplasm of a cow egg. One of the English MPs submitted, as you may read from the quotation, that in his soul he did not feel 80% mouse and 20% daffodil. Now, one of the classic reactions to this type of debates is to say that the English parliamentarian should go back to school or possibly go there for the first time to finally learn molecular biology, since he has not understood what it means that his genome, the totality of his genes, is correlated, linked by blood, to that of daffodils for 20% and to that of mice for 80%. This is one way of tackling this issue. I find another way, however, more productive, one that takes as its basis a different appearance of daffodils in English literature: this time around, not in the text of a parliamentarian debate, but rather in a poetry. Daffodils have been the subject of what is perhaps the most famous English

poem, *The Daffodils* (1804) by William Wordsworth. Wordsworth was thus another British gentleman who, while taking a stroll around the English countryside and seeing these flowers, developed a clearly anthropomorphic approach. He felt one with the flowers, as we may read in the last strophe:
> They flesh upon that inward eye
> Which is the bliss of solitude;
> And then my heart with pleasure fills,
> And dances with the daffodils.

A feeling of great proximity to this piece of natural matter, we might say, represented by the daffodils on that English hill.

Juxtaposing the parliamentarian who says, "I do not feel 80% mouse and 20% daffodil", and William Wordsworth, who instead felt one with these flowers, we might ask ourselves, as did another great English poet, where is the wisdom we have lost in knowledge and where is the knowledge we have lost in information? (Eliot, 1934) This is important precisely because information, understood as the result of digitization, which is the backbone of our age, is obviously integral to design students, who in design clearly adopt this digital result in various ways, including several biological ways as we will see later.

Eliot's words "Where is the wisdom we have lost in knowledge? Where is the knowledge we have lost in information?", applied to the daffodils, accordingly open to us a view, and we in fact recalled just a moment ago, with Foucault, how important it was to understand the type of look we have, since the moment we understand life through the genes and we understand genes as digital elements, we are also undertaking, in a vertical and somewhat swirling manner, this move from wisdom to knowledge and information, and we must be conscious of what these intellectual upheavals entail.

So much so that a few months ago this "tweet" from Genentech appeared, stating (on the anniversary of the discovery of the DNA): "Typing 60 words/min, 8 hrs/day, it'd take 50 yrs to type the human genome, that's 78 mil tweets."

I do not think I need to add anything else to prove more distinctly the utter conflation this "tweet" makes between two vast digital worlds, the world of life, reduced (and is not a pejorative term) to digital sequence, and the digital world that underlines Twitter, Facebook and the whole virtual world, yet utterly tangible in its effects, rotating around not only the social networks but, more generally, the digitization of our interpersonal lives.

As always, of course, the "tweet" has an author, and it is not an irrelevant author, given that it is Genentech, one of the most important bio-technological industries worldwide, undoubtedly the first one to embrace a vision which within the sphere of your studies should particularly concern you, i.e. turning biological knowledge into a product so as to design, in the real sense of the word, new products. It goes therefore without saying that the author of this "tweet" invariably carries along the dimension of capital. It is clear that design students who generate products in the XXI century should also familiarize themselves with the capital flows that characterize this century and, in this specific case, the way capital flows intercept life flows.

Now, keeping in mind digitization, ten years after the decoding of the human genome, we face two different images: the first which almost comes directly from that archaeology of medical knowledge of Foucaltian inspiration we recalled earlier, and the image of exhibited bowels, or the viscera of a body exhibited as they must have appeared to the field masters during a lesson of anatomy or anatomical pathology. The second, instead, is another type of image, a punctuated one, which divides the living matter into 23 chromosomes, where the many dots represent the signals coming from the sequences of the various genes contained within them.

A second juxtaposition is the one between two large spaces of knowledge: on the one hand, Padua's anatomical theatre, a theatre where the first dissections were taking place, later to transmigrate into that image Foucault defines as being a bit phantasmal, a bit ghostly of these organs that dissolve, come out, melt down, but also the place where a very small elite of persons, before a single expert who was the "field master", started seeing what we had inside.

On the other hand we find instead the Firefox browser onto which the human genome has been uploaded, which all of us and you can access by merely clicking UCSC and *zooming in* at the level of the single "base" or nucleotide, the famous ATCG, the "letters" of the DNA, whereupon we can eventually read, in the browser window, the exact sequence of each gene.

Evidently, much has changed from Padua's anatomical theatre which exhibited the viscera to a computer screen that exhibits the genes, but what is it exactly that has changed? Certainly, the depth of our look has changed exponentially compared to the one Foucault had already caught. If the slant of things, in fact, had already seemed clear-cut to him, i.e. the one that characterized the anatomical-pathological look that dissected the membranes of the brain and distinguished the brownish ones from the reddish ones, etc., the more vertiginous is our look nowadays, the neater it becomes now that it deepens and reads the ATCG.

How many are the genes in our body? 25-30.000, an approximate number, a figure perennially disputed, since it all depends on the way one counts them, but that is the order of magnitude.

The very fact that we may entertain this question and say 25-30.000, makes us realize that the depth of our look has undergone a major change. What has likewise changed, however, is the matter on which to formulate this look, i.e. the place in which these things are found, what they are made of. Because the viscera were a rather incontrovertible matter, and although a computer screen obviously has its own matter, in that case what interests us, of what this screen shows us, is not the matter through which we see it but rather the data, which as such might traverse a number of matters.

Of course, the places and protagonists of knowledge, too, have changed, e.g. one of the most important studies in the US, the dbGaP, is a huge database that links the genotypes to the phenotypes, all studies that attempt to understand which genes people have and how genetically diverse they are from each other and to link this to a whole series of features of their body and state of health, including their personality and behavioural traits, their predisposition to diseases, etc.

A huge accumulation of data that raises a series of very important ethical and political questions, so much so that not all these data are accessible to all and sundry, as a specific request is needed to gain access thereto. They are data that bring together many individuals, tens of thousands of individuals, and in future hundreds of thousands if not actually millions of persons, genetic data and very personal data, such as what these persons eat, what kind of sexual activity they engage in, if they had been criminals before, how tall they are, whether they are obese, etc. To come back to "matter", there are places where these data are kept, not incidentally called by a term that in English sounds as "treasure chests", which are isolated from any Internet connection precisely to avoid the "hacking" thereof.

You can thus see the significance of considering what is the matter on which we have exteriorized our genome, for one thing is to exteriorize it on the matter allowing anyone to click and see it on a screen, another thing is to exteriorize it on two large hard disks kept in a treasure chest and only accessible on request. The same type of data, therefore, journeys from the inside to the outside: keep in mind the movie trailer we have seen at the beginning; through the juxtaposition between the viscera exhibited in Padua's anatomical theatre and the nucleotides of our genome exhibited thanks to digital supports rendering them supremely immaterial as well, we have accordingly seen an externalisation process in a three-dimensional way.

By exteriorizing or externalizing the DNA, in fact (and always bear in mind the capital flows as well), the products, too, are exteriorized. This is what occurs in what has been so far the most important intellectual property, hence patent, case in the whole field of bio-technology, namely, the case in which an American firm, Myriad, patents the sequence of genes which, if mutated, put someone at risk of breast and ovary cancer. The patent triggers off from the opponents a legal battle that will reach the highest echelons of American case law and that will be settled, in 2013, by the US Supreme Court. The issue in dispute was ostensibly simple: can one patent a gene?

The judges had to settle the issue of the patentability of a DNA segment naturally present by virtue of the fact that it was brought outside, i.e. it was wrested from the context where it normally is, which is the nucleus of our cells. The Supreme Court ultimately decides that this is not possible, but that something else is possible, namely, to patent what is called cDNA (DNA copy). The cDNA actually is a molecule very similar to the DNA, if you recall your biology, even in broad terms, and if you recall that the genes codify for the proteins, in other words, they carry in their sequence the instructions to produce the proteins. This instruction, however, is not linear and immediate, in the sense that it is mediated by another molecule not incidentally called messenger RNA (mRNA), which the cell synthesizes by taking as mould the DNA, which is the one that will then be read to produce proteins.

This RNA is the intermediate stage making us shift from the instruction to the execution of the instruction, i.e. the proteins. Now, this intermediate RNA may be modified in a lab and transformed in turn into DNA, precisely this DNA copy, the cDNA. The fact is that, of course, if one thinks of the gene as information, the information contained in the DNA, once it has been transcribed into RNA, the information in the cDNA is exactly the same. However, the Supreme Court will say that one is not patentable, i.e. it cannot become a "product", whereas the other can. It does so through a series of passages and the Court states: a piece of DNA has a chemical property utterly different from the native DNA. In other words, once this single gene, who puts one at risk of breast cancer, is removed from its context, it has a different chemical property and is thus not patentable.

This of course attracts a multitude of criticisms from scientists as well as from jurisprudents, but the judges defend themselves and stress that several biologists and even several judges had stated that if this gene must issue instructions to produce a protein the dysfunction of which is important for breast cancer, the function of this gene, i.e. the informational content, is the same whether it lies inside or (by running a lab test) outside the cell: it is invariably the same thing.

The judges answer in the negative "The claimed isolated DNA molecules are distinct from their natural existence as portions of larger entities, and their informational content is irrelevant to that fact. We recognize that biologists may think of molecules in terms of their use, but genes are in fact materials having a chemical nature."

One thing, therefore, is to be matter along with 25-30.000 mates of theirs, and another thing to be matter far away from those 25-30.000. It is on this distinction that the question on patentability is based. In fact, they will then ultimately say, in actual fact the DNA per se may not be patented, whereas that molecule is patentable. Scientists allege that the cDNA may not be patented as the sequence is dictated by nature. Given that, however, the process is carried out in a lab, it is, in other words, a product, it accordingly follows that it may be patented.

To clarify the matter further, I compare the "messenger" RNA molecule with one of the most famous fragments of ancient times, the *Heraclitus Fragment*: "nature loves to hide itself."

With the idea, that is, of having to understand what is natural and what is the right point and the right scale for defining it: in holding that the DNA is natural and as such may not be patented, whereas the cDNA, while having the same information, is not susceptible of being patented, the US Supreme Court has found the solution to Heraclitus' question.

Many clinicians, however, rebel against that – let us go back to matter, internal and external, always remember to be on that tray – and say: if we clinicians oppose the monopoly of Myriad, which forces one to pay thousands of dollars for each genetic test on women potentially prone to breast cancer, what can we do?

The New York Times mentions the revolution carried out by American family doctors, who state that when these women undergo the test, it first acquires its own paper materiality that later becomes virtual, an e-mail in the e-mail address of family doctors. How can one then "bypass" the monopoly of Myriad, which wanted to keep the entire database of the tests secret? The solution put forward by the family doctors is simple, i.e. drawing this information from all the colleagues and thereby recovering the matter into which the sequence of that gene in woman x, y, z, etc. had been translated.

It is therefore important to understand what route is followed from inside to outside, for it is one thing to tread it through the Supreme Court, one thing to do it through the website of Myriad, and another thing to do it through border doctors who organize a *de facto* radical network to circumvent monopolies.

So much so that in the project that leads from the inside to the outside, more and more persons, the patients, take destiny into their hands. An example thereof is the saga of the American actress Angelina Jolie, who takes us back to the very concrete issue of the genomes, rather than that of the genome as abstract asset of human knowledge; the issue of the genome of the single individual, which in Angelina Jolie's case inspires

pretty radical life choices, i.e. the removal of the two breasts due to her discovery of the gene – for which Myriad had asked the patent – lest she develop breast cancer.
A very personal, very corporeal destiny, quite "inside matter", to return to Foucault. Another example of the extent to which genomics has become "personal" is represented by *23&me*, (23 understood as haploid kit of chromosomes), which established itself as the first company offering genetic tests directly at home in just four weeks. The inventor of *23&me* is the former wife of Sergey Brin (co-founder of Google) and here chances again upon that "conflation" between the digital element and the genetic one.

ACT II

The "molecular look", so vertiginously deeper than what Foucault had already introduced us to:
- Enables the visibility of life (genes, cells, molecules, neuronal circuits) that stand out from the context and are increasingly invested with agency;
- Generates epistemic tensions on the meaning of those units through observations that are more and more precise but more and more fragmented;
- Enables mobility/exchange/trading of the newly visible "units of life".

Why is all of this possible? It is possible because it is digital. Because it is a way of understanding, of studying life insofar as it is digitized. We have seen it at DNA level, we are currently seeing it at the two higher levels that are: *epigenome* and cells.

We carry out a very brief introduction to the *epigenome* from a schematic image, where we can see the double helix of the DNA, in which we also recognize the two letters "C" and "G", which gradually becomes condensed into these far bigger structures, and at the end what is seen summarized as if a cartoon is a chromosome, which in turn lies in the cell, and we see how this DNA is inhabited by "small men" involved in various activities. What did the Japanese artist want to convey in this graph?

He has actually summed up in a very attractive manner the essence of *epigenomics*: *epi* is a Greek prefix meaning "above", although there are many other definitions of *epigenomics*.

Epigenomics essentially studies the way our genes are regulated and the way they react to the environment. We have 25-30.000 genes, but are the genes of a brain cell and the genes of a liver cell the same or not? Do all our cells have the same genes? Depending on the giddiness of our look, the one more right is he who believes that all the cells have the same genes for that is how it is, though it then depends on what we mean when we say "same".

It is important in this regard to understand the reason why a neurone clearly is quite different from a skin cell, and has a very different form and function simply because it uses in different combinatorial ways the same 25-30.000 genes.

When a foetus is developing and starts producing brain on one side and skin on another side, each time these cells proliferate to "make the matter", to externalize that genetic program in a body, it is necessary to have what in scientific jargon we term "epigenetic memory".

Imagining that out of 30.000 genes 10.000 has been switched on and 20.000 turned off to generate a type of cell, this kind of information on the "use" of the genes needs to be somewhat stabilized so that the skin becomes and stays skin, and the brain becomes and stays skin.

This thing is intrinsically porous to the environment, where environment means what I eat, or what we are doing right now.

If, therefore, the strength of molecular biology is to digitize life through the DNA, *epigenomics* takes the next and more radical jump, i.e. to digitize environment. Because, if I can read through a digital profile the impact of environment on my genome, characterizing the epigenome and no longer the genome: observing, for instance, that two persons ate different things yesterday as we can read today on the stomach cells different imprints which yesterday's nutrients have dug into the way those 30.000 genes are regulated, epigenetics is a way of digitizing what seemed only a few years ago "un-digitizable" *per se*, i.e. the complexity of the environment, of the biography of what we term the idiosyncratically human:

"(...) its current and unifying thrust is, in a nutshell, the promise *to capture the analogical vastness of environmental signals* through the digital representation of their molecular responses. If what seemed irreducibly analogic (the social, the environment, the biographical, the idiosyncratically human) needs to be overlaid onto the digital genome of the informationally ripe age in a dyadic flow of reciprocal reactivity, the nit seems that this overlay can succeed only once the anagogic is interrogated, parsed and cast into *genome-friendly, code-compatible digital representation* (RNA, DNA found associated to specific chromatin modifications as in chromatin immunoprecipitation ChIP, methylated DNAs etc.)". (Meloni & Testa, 2014)

If, then, I read the genome as ATCG sequence, the *epigenome* is nothing more than reading the same thing to know whether in that cell, at that moment, that gene has been expressed at 1.000 or at 1 per hour, and whether this level of expression has been retained today, tomorrow, in one year or ten years' time: I have transformed the environmental input into a data I can read in a manner compatible with the genome and I can thus lay in those same "treasure chests" (in an even more problematic and interesting manner).

ACT III

Let us turn now to the last aspect of our *tour de force*, the cells that will make us understand why what we mentioned on epigenomics is absolutely essential for a comprehension of what I term the "biology of development". Louise Brown is the first human being born out of in vitro fertilization. It is a great moment of exteriorization: for the first time in the history of humanity, the human embryo, during the first days after fertilization, exists the womb of the woman who had copulated a short while before, and arrives at a laboratory plate, turning into a scientific object, and soon into a commercial object as well, a modifiable object, perhaps one of the most striking examples of "internal-external" itinerary. That embryo, once we have taken it out, knowing how to take it out, can be put back in by us. In Louise Brown's case, in 1978, it is the embryo that is externalized from a viewpoint of generation, i.e. to produce other human beings. In 1998, however, the externalisation of life takes a very different path, that of "regeneration". Dolly, the Scottish sheep, is the first cloned mammal: an animal born out of a skin cell. Nowadays, in fact, we know that this skin cell, as skin, had switched off 20.000 of those

30.000 genes, and it was not known if such a state would be epigenetic, i.e. if it only carried the memory of being "skin".

The moment however that such a DNA "wrapped" in those chromosomes that silenced 20.000 genes and activated the other 10.000 in order to be skin is implanted again in a cell previously deprived of the nucleus, is this epigenetic configuration still capable of letting me recreate an entire organism from scratch? The answer, quite remarkably, was yes.

The "embryonic stem" cells are those cells that "captured" and externalized again by the embryo in a precocious phase, are capable of giving birth to all the tissues of our body.

The idea of "regeneration" derives from the possibility that if from a person I take the nucleus and I implant it in an egg, this egg "reprograms" that nucleus. If, for instance, that nucleus used to impart instructions to be skin to that cell, the egg is capable of reprogramming it and making the whole development start afresh. This means that it is possible to create embryonic stem cells from which we may produce various tissues sharing the same DNA of the person from whom everything began, and that, accordingly, it is possible to conceive the idea of repairing/transplanting organs or tissues without running any risk of rejection.

The aforementioned vision is also defined as "therapeutic cloning", which so far has not ever taken off as it is a very complex technique. Only ten or fifteen persons throughout the world is capable of bringing off the "little game" of the egg.

It is precisely the "artisanal character" of this process that becomes a decisive factor, given that design and craftsmanship capture different moments of making the things: the individual scale versus the industrial scale. In 2006, however, a Japanese colleague succeeded in his industrial evolution, and was able to ensure that the reprogramming of the adult cell into a "pluripotent" cell (from the Greek *pluri* -many, power), one capable, in other words, of generating all the tissues of our body, could be achieved in a very simple manner, at least for a good molecular biology lab. What he accomplished was to take the cell of the skin and of the 30.000 genes, force the expression of merely four of them, four strong genes we might say, that are able to reprogram the epigenome by changing the identity of this cell (which only knew how to be skin), making her start over from scratch "as if" we had carried out a nucleus transplant, though in a much more accessible way.

Today it is accordingly possible to speak of cells "to order", so much so that the first pioneers in the use of this technology speak of "Cell Fate Plug and Play". (Chambers & Studer, 2011).

The identity of the cell is by now a combinatorial game of plug and play, since use is made of the four skin genes to obtain a pluripotent cell, although in actual fact, by taking three more from the skin cell, we may directly obtain a neuron or a piece of heart or muscle, etc. The domain of combinatoriality we have observed at the genome level has been brought back to the cell level, thereby occasioning a huge jump forward scale-wise.

What are, however, the implications of all of this? They are implications that oscillate between two visions, two great myths that, obviously with a touch of lightness, I have associated with two recent Hollywood movies: *Prometheus* and *Avatar*.

Prometheus is by definition the myth of regeneration: being sentenced to have the liver devoured and then growing back was on the one hand an image of eternal pain, but, on the other hand, a great promise of rescue as well, the idea of a possible regeneration. The other example is Avatar, the idea of generating representations of ourselves; in the movie, the *avatar* was a kind of shell that reproduced the real person, moving around and acting in his stead, an example par excellence of the trailer that moves from the inside to the outside.

In which sense the technology I am speaking about makes *avatars*? Because today it is really possible to move from a person affected by any type of genetic disease, reprogram his skin cells (though nowadays it is already possible to do so from his blood cells, from a strand of hair, even from the cells we leave behind in the urine), bring them back to a pluripotent state (PS, stem cells, induced pluripotent stem cells) and then differentiate them in vitro, transforming them into neurons, muscle or liver cells, etc. Cells, moreover, are not only generated to be inserted inside the body in order to repair organs, but also to study diseases, to test new drugs, to do whatever had never been possible in the prior medical history (we go back to Foucault), i.e. to explore the history of the disease in a lab.

In fact, the doctor who, in the second quotation from Foucault, described the "membranes of the hard mother" of the brain, was of course observing these things on a cadaver, and even if he succeeded in getting "the clear-cut slant of things", the history of this disease was hindered from him, as it was like getting to the scene of the crime after it had already been perpetrated.

To better understand these issues, I am citing here some examples drawn from a research work I am conducting with a student of mine, which I have precisely called "This is avatars" (and which recently won a European GRANT), i.e. the *avatar* of disease. It is thus possible to conduct tests on a tissue that finally makes sense, being a sick tissue, which for obvious reasons could not hitherto be taken from the person, and being above all a sick tissue at the right time of onset of the disease, I have externalized its history.

I quote a series of phrases taken from the most important articles that are advancing this biology frontier (which in many respects is a design frontier as well): "iPSC SOD have been made from a symptomatic 82-years old patient and differentiated into motor neurons and glia (Dimos et al., 2008), but it is not clear yet whether these cells are symptomatic in any way" (Colman & Dreesen, 2009).

Foucault would have a lot to say on this semantic upheaval, whereby the symptoms of a person unable to walk are molecularized and we think we might discover them at epigenome level, i.e. at the level of which of those 30.000 genes are switched on or off in an in vitro, let us say symptomatic, cell. We speak in fact of this field as the study of diseases "in the dish", and many persons refer to this as "the new patient" the patient of the XXI century.

In the pipeline there is also the 3D printing of organs; being in fact capable of extracting the digital code of cell, we are able to modify their identity, so we may start becoming ambitious and think of assembling them into organs or at least tissues. A "mini-brain" or a "big brain" (Lancaster et al. 2013), is the result of a skin cell reprogrammed into a pluripotent cell and grown in vitro in such a manner that, alone, the cells might be

capable of assembling a structure with an amazing similarity to the three-dimensional structure of a human brain.
Another example of the intellectual ambition arose during the conference we organized among students, which we called *"From genes to thoughts"* (EMBL PhD Student Symposium on Neurobiology, October 20th-21st, 2000 at FMBi, Hedelberg, Germany), within the scope of which I fell in love (intellectually speaking) with a very peculiar disease termed Williams syndrome. It is a disease where children have a form of mental disability linked to two very specific traits: hyper-sociability, being also defined as children with a "cocktail party personality"; compared to children affected by the Down syndrome, who have a similar level of intellectual disability, children with the Williams syndrome have a very good language, are strong in language, in the description even somewhat baroque of things, but are very weak in what we term visual-spatial cognition. They are children who lack 26 of those 30.000 genes, and, moreover, of these 26 they only lack one of the two pairs (we have two pairs of nearly each one of our genes); these children lack the 26 pairs of the mother or the father, hence it is one of those sicknesses we call "dose-related", a slight difference that produces this phenotypic level. A few years ago, it was also discovered that if the dose changes in the opposite direction, i.e. it doubles, and of these 26 genes, we get three or four pairs, a form of autism and a serious linguistic deficit arise. We are talking of a few genes and of minimum differences in dosage, which nevertheless seem to control in a symmetrically opposite manner two essential aspects of the human condition: sociality and language.
My lab is studying this type of problems through reprogramming, by taking cells from the skins of these children and reprogramming them, we have produced in vitro neurons.
A recent article, *Toward the functional annotation of human genomes: sequencing meets cell reprogramming,* deals with the proposal to create "banks" in which to include 10.000 European individuals whose cells are reprogrammed: think of what a scale of externalization that is!
As Ian Wilmut (Dolly's "father") wrote, "We have entered the age of biological control", and control is of course an essential element of design.
It is no coincidence that another work under preparation with a student of mine, titled *"Scales in action: competing epistemologies and innovation regimes"* (NYSCF v. Roche), analyses the new large-scale frontiers of pharmaceutical industry that talks about using these new techniques.
The work aims to understand the type of epistemology, of philosophy, that underpins this type of work, and I quote a colleague who mentions how these cells are produced:
"Right now, we make the nice stuff, we make iPS cells like Louis Vuitton hand bags. But the field is moving to the big stuff, and if you want to be part of it, if you are going to make thousands and thousands of iPSCs, you have to do a more scalable production, and you need to automate. You need to be like Wal-Mart, if you want to be relevant (...)". (Chad Cowan, iPS Core Leader)
In actual fact, in my lab seven are the persons who still work in a rather small-scale fashion to produce the cells of these children affected by the Williams syndrome (WS). This explicitly makes you realize the extent to which bio-technology has nowadays succeeded in externalizing life, the vital information, in order to later re-internalize

it, capable as it is of pursuing various levels of cellular engineering and design that accompany the entire "range" from Vuitton to Wal-Mart.

"(...) Imagine, in the long run, that it will become routine not only to access the complete genetic information of a patient but to directly probe the patient's own iPSC-derived tissues for a broad range of medical questions". (Lee & Studer 2010, p.27)

"iPS cells can be generated from any human who is taking a medicine. Thus, effect or lack of effect of a particular drug that is detected during clinical treatment can be reanalysed using iPS cells from patients". (Nishikawa, Goldstein, Nierras 2008, p.727)

CONCLUSION
Will we ever be able to do what an egg is capable of doing, "playing" with the genes? A few years ago, nearly all the scientists answered in the negative, yet not only is it possible, but from 2006 to 2014, within the space of eight years, we are contemplating the idea of moving from Vuitton to Wal-Mart.
I would like in fact to conclude with the idea of what externalization means. When we externalize, we do not only externalize things, but also values, epistemology, regulatory needs.
The biologist-entrepreneur Creg Benter who takes the gamble of sequencing the entire human genome in one year with private money has gone quite far, and a few years ago alleged that he had produced the first form of synthetic life: he produced a bacterium with the DNA entirely synthesized in a lab. The frontier is that of the so-called synthetic biology, the one somehow programmatically closer to the world of design. So similar was this organism that it has to be labelled, have "watermarks", i.e. some small DNA pieces, to make it patentable, as it would have otherwise been impossible to distinguish the natural from the synthetic. It is the moment in which it is possible to write in the DNA a patentable rationality: a commercial interest that turns into matter within the living matter.
The last example again sees America as ground for discussion. During the years of Bush's presidency, embryonic stem cells and cloning were very controversial topics. Bush's bio-ethics committee alleged that we need embryonic stem cells and that it would have been a great revolution getting them through the cloning process, even though to do that an embryo, with all the attendant ethical problems, is created.
The committee accordingly turned to one of the most prestigious genetic engineering labs to try and avert the problem. The solution was that, prior to commencing the process, a gene responsible for letting the organism develop fully is eliminated. How can we possibly know, however, that such a gene has been removed? It has been inserted close to another gene that emanates green light when it is irradiated by blue light, hence the green embryos will be those most morally neutral, least ethically problematic. One of the greatest neurobiologists, who was part of that committee and strongly opposed this proposal, speaks of *logos* and *bios*, and says: "Normally we generate a word to describe a biological phenomenon, and here we seem to be tinkering with biological phenomenon to have it fit the meaning of a word" (Michael Gazzaniga, 2004).

In other words, we are playing with a biological phenomenon to make it acquire the meaning of a word, put it differently, he is stating quite concretely that *logos* is coming into *bios*.

What is evinced by these two examples is how, in different fields, the explicit discretion of regulatory demands and political-commercial interests inside life, hence inside design per se, has led to the definition of "bio-constitutional moments". The generation of new forms of life inside the DNA will also carry the imprint of the new political demands: a moment of constitution, albeit one not necessarily written into the text of the law, is unfolding itself.

References

Chambers, S. M. & Studer, L. (2011, June). Cell Fate Plug and Play: Direct Reprogramming and Induced Pluripotency. *Cell 145, 6*, 827-830.

Colman, A., Dreesen, O. (2009, September). Pluripotent stem cells and disease modeling. *Cell Stem Cell 5(3)*, 244-7.

Eliot, T. S. (1934). *Cori da "La rocca"* Milan: Rizzoli.

Foucault, M. (1998). *Nascita della clinica. Una archeologia dello sguardo medico*. Einaudi.

Lancaster, M. A., Renner, M., Martin, C. A., Wenzel, D., Bicknell, L. S., Hurles, M. E., Homfray, T., Penninger, J. M., Jackson, A. P., Knoblich, J. A. (2013, 19 September). Cerebral organoids model human brain development and microcephaly. *Nature 501*, 373-379.

Lee, G. & Studer, L. (2010). Induced pluripotent stem cell technology for the study of human disease. *Nature Methods 7*, 25 – 27.

Meloni, M. & Testa, G. (2014, August). Scrutinizing the Epigenetic Revolution. *Biosocieties 9*, 431-456.

Nishikawa S., Goldstein R.A., Nierras C.R. (2008, Semptember). The promise of human induced pluripotent stem cells for research and therapy. *Nat Rev Mol Cell Biol. 9*, 725-9.

Nowotny, H. & Testa, G. (2011). *Naked Genes. Reinventing the Human in the Molecular Age*. Cambridge, Massachusetts: MIT Press.

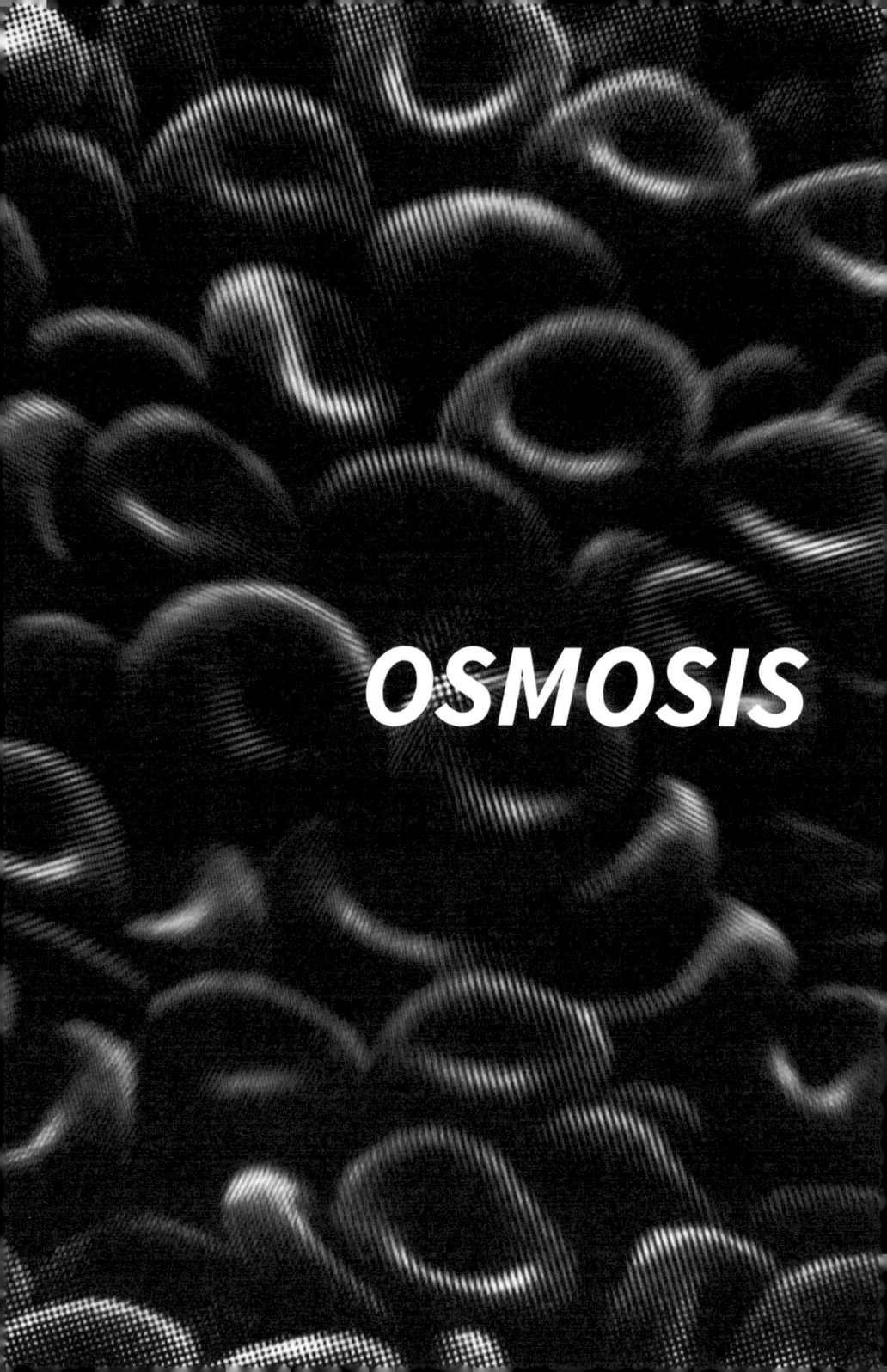

BALANCE BETWEEN THE INTERIOR AND THE EXTERIOR

MASSIMO FACCHINETTI
Facchinetti & Partners, Bergamo, Italy

Osmosis as an in-between element, a joint. Unexpected comparisons on a variety of themes have led to the reinterpretation of the concept of osmosis. Starting with art, together with Michelangelo Pistoletto, we can view artwork under a new guise in which it looks towards the future to then look back over the past; we continue with Clino Trini Castelli experimenting the ability to look at material three-dimensionally and to lead it to the dimension of space between interior and exterior; we then move onto the experiences of Roberto Cingolani who extracts his translation of the qualities of biotechnology from the functions of the human body; and we finally reach Chris Bangle who interprets design in movement, in its immediate transformation. We reach a perception of infinite osmosis between different disciplines and their importance for the future of design. The complexity and diversity of experiences and themes confer a different essence to the concept of osmosis.
The interdisciplinary character between art and design, concept design and biotechnological engineering applied to the complex machine of the human body has led us to look beyond the solutions established by the knowledge of a designer.
It is increasingly important to know how to interpret the human body, to know its limits and imitate its qualities even while designing living areas, with devices that simulate such characteristics. From the beginning of time, research into what our body can express has represented a continual research of man himself: something that today is also explored through an approach that involved different meanings than the traditional methods of medicine: from genetics to the ethical and spiritual dimension.
The experience lived in the "osmosis" section has clearly highlighted the possibility of designing living areas while integrating themes that had never been integrated before and not even seen within the aspect of qualities that the human body may suggest.
Osmosis as the act of interpreting, looking out to see within, such as the exterior skin of a building or environment, which is no longer an element separating the exterior from the interior or vice versa, but which becomes an element that unites and allows us to perceive with all five senses, to see and communicate what is inside or feel and breath in the area between interior and exterior, to listen to and touch what changes in the function of a final goal.

Osmosis as an element of exchange between interior and exterior, to understand what is in, through membranes that may allow this penetration and the passage of sensoriality, from the air to temperature, from sight to touch, from taste to smell, but also the perception of the immensely large compared to the micron.

The interpretation given by various experts allows us to understand how we can look beyond that which has already been seen; entering, penetrating, using reciprocal influences to fuse in osmosis the point of balance between interior and exterior, allowing students on the course to express themselves with a new concept of limitless space.

THE THIRD PARADISE

MICHELANGELO PISTOLETTO
Fondazione Cittadellarte, Biella, Italy

The Third Paradise is the current stage of my artistic journey (Pistoletto, 2013).
I have chosen the term Paradise without alluding to any religious significance, but rather for its primary, original ancient Persian meaning of garden.
The "first paradise" – according to the meaning common to a variety of religions – was that in which humankind was completely integrated with nature. Artificial intelligence had not yet been invented so this paradise was one in which humankind had no responsibility. The Persian's paradise, on the other hand, was the "protected garden". The Persians lived in the desert, where in nature there is no water, and so they had to artificially recreate an inhabitable nature. To do so, the Persians ingeniously created circular walls which allowed dew – humidity – to accumulate, meaning greenery could be cultivated therein. This device – the artificial ability to intervene in nature – gave birth to the "second paradise".
The "second paradise" marked the beginning of humankind's ability to leave nature to return to it in a new and different way.
Mankind has gradually evolved the second paradise until today, a time in which artificial devices have become all-encompassing. This passage is linked to the religious phenomenon of the apple as forbidden fruit: in the beginning, we were "in" the apple; afterwards, the bite marked the beginning of the second paradise and we "left" the apple, and now we have become completely artificial. In principle, the apple was the symbol of our relationship with nature (also linked to sex and, therefore, sin). Now, if we think of the *Apple* corporation the symbol of the apple has become completely artificial: a bitten apple where nature is no more.
Therefore, there is the first paradise which is natural, and the second one which is artificial.
Today we are witnessing a time of huge scientific, technological, economic and productive progress which has, on the flip side, caused an increasing deterioration manifested in a crisis that we have brought upon ourselves.
The symbol of the *Third Paradise* [Fig.1] unites the artificial with technology. This symbol comes from the mathematical sign for infinity which is a line crossing over itself to form two circles. The symbol of the Third Paradise is different because the line, crossing over itself twice, merely passes from one situation to another continuously: after a long journey, it returns to cross the line again, giving it a similar appearance to the image of the present in the mirror.

Fig. 1 Michelangelo Pistoletto, Terzo Paradiso.

1. ORIGIN

My artistic journey started with representation, using a mirror.
From the self-portrait through re-representation, I arrived at the mirror. Without this tool an artist cannot create a self-portrait, though he could also do it with a photograph. Thus I, looking at myself in the mirror, decided to create a self-portrait on the mirror itself.
Facing the mirror while represented on the same, I was within the mirror in a virtual area where reality, the artistic surface and the virtual space were connected. I concentrated more and more on the mirror itself, looking at myself as a figurative element, as it is clear that the mirror is depictive, not abstract, as it depicts everything that is in front of it. Looking in a mirror you can understand that images are not always the same – quite the contrary, an image seen once is not the same a moment later. The present is, in fact, this very crossroads, at point at which everything happens and stops at the same time, even the very concept of life and death: something that isn't there a moment before is not yet born; something that is not there a moment later is dead. We live this phenomenon of transition in continuation: this is an important discovery that I made while looking in the mirror.

However, we experience a different situation of continual transition as we live within the duration of it: we last and we are physically set up to last over a certain period of time; so just as an insect has its time, so too does a person, a society, an era, a civilisation.
For this reason, I changed the symbol of infinity so the line crossed itself twice in order to form a third circle between the two initial ones: the third circle is the duration, the period, the space in which time does not pass on immediately. Between the two circles of infinity is the third circle which is *finite*, the duration of life.
As I said before, in the symbol of the Third Paradise there are two outer circles which I have defined Nature and Technology. In the centre there is the "place of combination", also known as *osmosis*. The central place is the osmotic place of the two elements that we compare as they are already themselves activated in the comparison and exist as comparison.
The desire behind this sign came from the idea that a life, connected to the concept of good living, must in some way be defined, understood and communicated so as to be able to develop this circle not individually but as a part of society. A process that may be defined as a universal architectural project.

2. MIRRORED PAINTINGS
The beginning of my journey is marked by the self-portrait, or rather the search for identity. I realised early on that my identity could not merely consist of my personal image (eyes, face, etc.) but that I had to find my identity beyond my image.
When I was young, I had moments when it was very difficult to have an identity: I had to be Catholic and Fascist because they taught me to believe in God and Mussolini in equal measure. Then in 1940, when I was eight years old, war broke out. 1943 was catastrophic with the arrival of the American bombs.
This made me realise that I had to somehow leave the type of culture that I had been indoctrinated with by the eco-socio cultural systems. I understood that I had to search for my identity, and I was lucky enough to meet a great advertising designer, Armando Testa, who introduced me to modern art as that was the direction that advertising took. Advertising is very similar to design: it connects creativity to products through the market and mass communication looking at modern art from an aesthetic viewpoint. Advertisers "stole" and copied modern art for new aesthetic ideas to distribute and use. One day, I saw a piece by Lucio Fontana, a white "perforated" painting in a display window in Turin. At the time, everyone called it scandalous. I, on the other hand, thought "If this man has created this, there must be a reason for it; the problem is that I don't understand it because I don't understand his reasoning behind it, so I must find my own reason".
Modern art showed me that I could find my identity and this phenomenon did not isolate me from the world – quite the contrary, it threw me into the world with all its problems. As Italian art tradition is figurative and not abstract, I started thinking how I needed to start my work from myself and see how my image united with the image of the world. In that moment, I understood that modern and contemporary art had led every artist to the highest possible individual freedom, to the extreme ability of taking on all possible symbols imaginable within himself. A single symbol - his own - could represent all symbols: it was the exclusion of himself from all social

phenomena, absorbing the concentration and meanings and energy of society. An incredible freedom that was taught to me through modern and contemporary art. As was the case of Giuseppe Capogrossi, an artist who painted "forked" symbols; he multiplied and repeated them. I understood that the expressive self meant maximum freedom. But, as everything that I knew about contemporary art had already been done, I asked myself what I could do.

Slowly I understood that I would have to reach something that didn't represent "self", but "us". How can "we" find ourselves in this civilisation? and how can we meet ourselves? I started working on the relationship between the painted image and the mirrored surface: painted image has become photographic image, because the figures visible within the mirror are absolutely subjective, they cannot be interpreted. What the mirror says is the truth.

I could not represent my figure, my sign, my face, by painting myself in my style; I had to make the fixed image coherent with all the moving images that came within the mirror. So the choice leant towards photography because it captures that famous transition between life and death - it sets it - and that set image, inserted within the mirrored painting, becomes absolute stability, fixedness, together with the maximum dynamic: they are two extremes.

And here again we see the sign of the *Third Paradise*: maximum immobility with maximum dynamicity. The image that is fixed in a photograph immediately becomes memory, the moment in which it is photographed and fixed is already memory: it is no longer present and becomes the past.

In the sign of the *Third Paradise*, the past and present coincide in an osmotic way. [Fig.2] The *Third Paradise* is already a consequence of that mirroring phenomenon in which the two extremes – dynamicity and immobility of time – coexist.

An absolute painting, in which there are no positive or negative personal sentiments or emotions; but it is the work itself that shows how things are, becoming phenomenological, therefore scientific, piece where art and science "osmotise". Together with science, there are philosophical and mathematical references regarding ways of living: self and infinity, space and the universe.

The sign of the *Third Paradise*, as well as representing nature and artificial, represents all the opposites that may be "osmotised", that is interconnected within themselves, like an electric battery: the positive and negative poles in the centre produce energy. This is therefore also a symbol of energy.

3. TRINAMIC

Trinamic is the dynamic of the number "3".

The number three is not merely the sum of 2+1 or 4-1; it is a universal phenomenon in itself. It is the relationship, union, combination and osmosis of two elements that are separate and even opposite or conflicting. Two elements put together create a third element that didn't exist before. This is the phenomenon of creation: the combination of two units that give life to a third distinct and brand new one which is in the middle. In the same way as in chemistry where water is created by uniting oxygen and hydrogen, and in physics where heat and cold generate lightning (energy), and even in biology where a male and a female give life to a whole new individual.

Fig. 2 Drawing of the Third Paradise by Michelangelo Pistoletto.

Now, considering the social sphere, we can recreate the same phenomenon of *trinamics* with the Third Paradise: the sign of the Third Paradise is a *trinamic* symbol.
Dealing with the specific matter of the social sphere, considering the importance we give it, we must pay a certain attention to the contrasting elements. We need to understand the opposite of each word/phenomenon and try to find the connection so as to reach creation. We can create the new world by uniting all those contrasting, conflicting elements that seem insurmountable and which on the other hand, put together in a certain way, become what is new in society.

4. RESPONSIBILITY

Another fundamental value is that of responsibility. Those artists who left their mark and brought it to a level of subjectivity, complete concentration and expression, also acknowledged their great responsibility. They had taken on the responsibility of the mark, a responsibility that they had not been able to put into action as they had stopped at freedom. Freedom is personal, while responsibility is communal.
I am responsible for my sign, I have the greatest freedom, but when I find myself immersed in infinite freedom, what happens? I can use it in a positive or negative sense, create it or destroy it. Freedom in itself is not sufficient. This is why I turned to "us": I realised I wasn't alone in the mirror; I was essential because without my eyes, without myself, the mirror would not work, it wouldn't exist. But in reality, the mirror exists because everyone's eyes make the mirror work. I saw how, inside my picture, inside my piece of art, everyone existed, thus bringing out a subjectivity of representation and at the same time a distribution of that freedom that was not merely of the central object alone but of all the objects that entered the mirror's range. I felt that freedom became a giant wave that led me to indicate responsibility, because my freedom was no longer sufficient.
Freedom must be individual but, in order to continue being so, it must become social responsibility. For me, it seems that freedom and responsibility are the two points that meet to create society within themselves.

The sign of the *Third Paradise* is also a set of scales: we need to rebalance individual freedom with responsibility. Individual and social freedoms together produce a balanced society within.

5. CONCLUSION

We live in a very significant scientific and technological time. Today, we have the ability to manipulate DNA and this discovery is even more outstanding than splitting the atom. Splitting the atom led to the creation of the atom bomb, with the consequent possibility of destroying millions of lives at once. Now what will be possible with these new discoveries?

Cittadellarte is the place where we try to bring aesthetics and ethics together.

It isn't important to merely look at all great innovations, which are all of great interest; we also need to understand what they are like and how they can be used.

We need to calibrate a way of using this science. This means that today we must develop a sense of social ethics, social morals at the very centre of this design; not the old "moralistic" morals, but morals that really take into consideration the infinitely small and the infinitely large, which considers this enormous power that human beings possess. We are monkeys with computers. Mentally we are very backward: we must start, if not from zero, at least from three, from the Third Paradise. We have to consider everything again: the very phenomenon of spirituality, as guided and monopolised by religions which are conflicted and a fundamental tool of the mentality of war. We need to start sieving through the past to understand what may be done for the future. This is the Third Paradise.

If we don't put all of this into practice, the planet and humankind will self-destruct thanks to the economic crisis, consumer growth and over population. We need to try and reach a balance in economy and global growth. We need to get moving.

Fig. 3 Reintegraded Apple by Michelangelo Pistoletto, Piazza del Duomo, Milan 2016

References

Pistoletto, M. (2013). *Il Terzo Paradiso*. Venice: Marsilio.

MATER MATERIA

CLINO TRINI CASTELLI

Castelli Design, Milan, Italy

Before coming to the day's theme, I would like to say that the documentation that you are about to see is so vast that I have decided to limit it to the thirty years between the 1960s and the 1990s. Amongst the terms and definitions that will appear in this presentation, there is one that particularly stands out, though that has been the case only for the past decade or so, or rather since I focussed on the situation during the history of my work in the field of colour, materials and finishes: it is the acronym CMF Design, which has today become a familiar term. I am especially pleased with the fact that I was able to develop this theme while keeping it within the culture of design. For the most part the theme of CMF Design – with all its emotional implications – has been kept within the world of style and fashion: a reality made up of transcendent visions, far from the typically immanent outlooks of design and its culture. All of this has been possible thanks to a series of personal coincidences and stories that I will tell you about today. I believe, however, that my most original contribution in this field was that of having introduced the meta-project to CMF Design, leading me to a systemic vision that was particularly suitable for perceiving the qualitative dimension. This approach has then been utilised over the past thirty years by people working in the industrial field on themes linked to emotional factors. To give an example, of all the production areas of the automotive sector the most expensive and complicated is painting – colour, in short. For cars, in fact, the hardest productive investment to optimise is that of the emotional characterisation of the product. So it is easy to understand why, when I first spoke of how to implement the emotional identity of the products by rationalising them, the companies in the automotive sector – first the Japanese, closely followed by the Americans – reacted very positively.

And here we come to the themes of this lecture: materials and the innovation of plastic languages. This is a dimension I had already dealt with in the 1960s, but which then spread and became more established from the 1970s to the 1990s. We will see how it has become so important, even concerning the strong seduction of shapes. My goal is to understand the importance of materials as original and new matrixes, major figurative reference points of the nature of objects. If we succeed, this would be a huge step in the right direction. When, at the age of 17, I moved from the Fiat Central School of Turin to take over the position that Giugiaro had held six months prior, I discovered the image of a detail of a product: Olivetti's Elea calculator designed by Ettore Sottsass. Here in fact I had free access to all design magazines and, looking through a 1962 issue of *Domus* I discovered this incredible image in that seemed like a new and liberating dimension that could free me from a kind of obsession. For years, understanding the origin and the reasons for that particular attraction felt like a constant need. But we will return to this aspect further on.

One of the drawings I made to be accepted in the Style Centre in 1961 [Fig.1] was an exercise in perspective that represented a computer room: do not ask me how or why, but the fact remains that I was already moving towards the design of systems for data centres, which I would later do all over the world, and for the rest of my life. This was in fact the Elea, a wonderful machine that had a quality that for some reason fascinated me, which a few decades later I would define as "No-form". To the right of the calculator you can see a plaque. At the time, Sottsass took me from the Style Centre to the Milan-based Olivetti studio. Considering my specific interest, I asked to develop the design of that plaque which, even if belatedly, brought me into direct contact with that project. My work with Ettore Sottsass's Design Studio for Olivetti would deal with the interfaces of all the products, an interesting activity in a fantastic environment, a very different world from the automotive one I had just left behind.

Fig.1 Clino Castelli, *Data Center*, 1961. Courtesy Clino Castelli Archives.

These two images [Fig.2-3] are also important because they regard my social life at the time. I was friends with Alighiero Boetti, Michelangelo Pistoletto, Piero Gilardi… What you can see are two portraits: the first is by Pistoletto, made in 1963 when I was 19 years old. The second is from 1967 and Boetti did it, at the time of his first *Arte Povera* exhibition. I have shown you these two images to confess that I was undecided about whether to remain in the world of design or to enter the circle of friends and galleries, where I was able to work in a strictly artistic way. I must say that the fact that

Fig.2 Michelangelo Pistoletto, *Ritratto di Clino*, 1963. Courtesy Clino Castelli Archives.

Fig.3 Alighiero Boetti, Clino, 1966. Courtesy Galleria Christian Stein.

I ended up in Milan with Sottsass encouraged me to continue working in design, also due to the prospects that started opening up for me. But there was another aspect that concerned me more, of which I have become completely aware only recently: I was certainly interested in art, but above all I was searching for innovation in plastic languages. In fact, one thing that I love and which I believe I am good at is searching for innovation in things, and therefore what interests me in art is the innovation of expression and the various forms of figuration. Unfortunately, I think that most of what has happened of importance in the world of art, starting from the mid-1960s, regards other types of innovation, those regarding forms of behaviour, which still today do not show any sign of changing. Human behaviour certainly creates great pathos, but there is a problem: it does not create plastic innovation in figurative languages. I am thinking of the shock of those who saw the very first exhibition of the Impressionists, the use of the new technique of perceived colour, of partitive synthesis. Works of art can represent something else, unlike behavioural art – Marina Abramovic's performances, for example – where you will find a forceful media-based elegance of languages, but no true plastic innovation. Alighiero Boetti, for instance, was a huge innovator for contemporary lettering, producing his famous maps in Afghanistan. The weavers used the characters of the Latin alphabet without understanding their meaning, thereby transforming them into simple icons, abstract coloured marks. In this case, I sense a new vibration, a change in graphic languages that allows me to differentiate, for example, between the Sixties, the Eighties or the Nineties and so on. The alternation of major figurative styles remains one of the things I find most fascinating in the history of art. Unfortunately, the last time we had any major innovations was the period between Minimalism and just before Pop Art and Arte Povera.

In 2014 my pieces *Vuoti Ceramici* [Fig.4] were exhibited in the Milan Triennale. I made them in 1964 when I was twenty years old and had just started working for Sottsass. They are ceramics that already had a non-compositional approach, since my sole interest was the expression of the material. This "Ceramic Void" was designed like simple metal tins, objects without formal research apart from the openings created to empty the containers. These ceramics clearly show that even then I had no sympathy for traditional plastic research (Sottsass understood this attitude immediately, even if he defined it later as "pop"). I had graduated from the Style Centre, where I was trained to design automobile bodywork, but without being seduced by those forms: a vision that was already part of me and which I would later define as *No-form*.

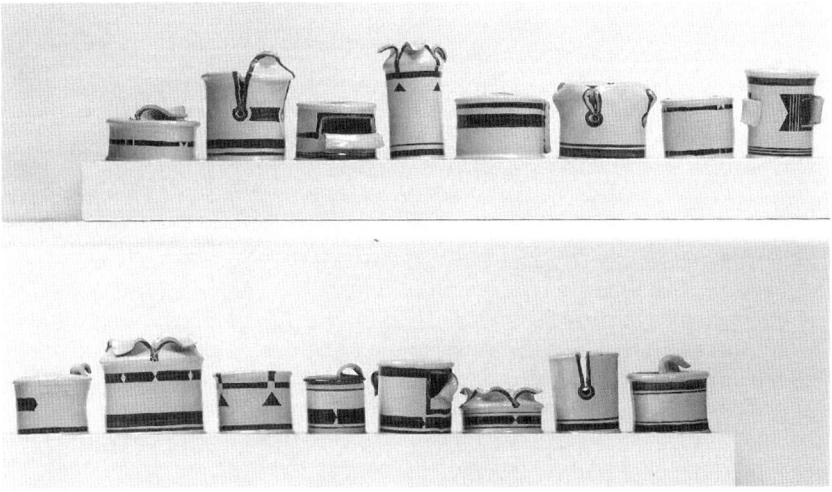

Fig.4 Clino T. Castelli, Vuoti Ceramici, 1964. Courtesy Clino Castelli Archives.

With the poster I designed for Galleria Sperone in 1967 [Fig.5], I tell the story of another important meeting, another event that had a great influence on me. The story concerns my discovery of the work of Dan Flavin, which for me was a memorable event, because for the first time I understood what it meant to deal with new figurative realities working with an absence of form. These works were created with coloured fluorescent tubes, objects that already existed but whose light became a figurative expression, a vibration of pure energy. Here the compositional aspects did not come into play at all. There was no formal research in the neoplastic sense, just as there was no traditional research into harmony or beauty. The result, however, was extraordinary: it was a new

Fig.5 Clino T. Castelli, Dan Flavin poster, Galleria Sperone, 1967. Courtesy Clino Castelli Archives.

language in which I saw something truly different, which fascinated me more than everything else that, though beautiful and interesting, could be found in art of the time. The fact that I did not use Sottsass's style to imitate him, as the others in his studio did, was something that piqued his curiosity. This is perhaps why he asked me to design an object for his new house on via Manzoni: a wall lamp to mount on the furniture he used to divide his living room. I created a small series of self-produced lamps from a single ribbon of sheet metal that curved away and then returned to its base [Fig.6]. This meant the light could be turned up or down, once again without particular research on form. I only understood the real reason behind my vision of that design a few years ago. I had enrolled in the Fiat Central School at the age of fourteen, where I spent the first three years learning techniques for building automotive moulds. Do you have any idea what that job was like, in an age without computers? It meant having to create objects exclusively by hand, tracing complex curves along scratched marks and estimating tolerances by sight, to within a tenth of a millimetre. We worked with lines marked with little crosses, which within the volume of the steel mould defined an ideal solid that was then extended. All of this was done by hand, and any mistake led to huge problems. It was the world of trigonometry, of hundreds of square roots extracted without a calculator… the most tedious thing in the world! If you were to put a young man in those conditions today he would probably lose his mind. In fact, I believe I did lose my

Fig.6 Clino T. Castelli, Noform Sconces, Casa Sottsass, 1965.
Courtesy Clino Castelli Archives.

mind and subconsciously promised myself that I would never again do something that had a form. This old photo shows me in the Fiat Style Centre, where I was junior stylist: I had finally escaped from the "mould tracing hell-hole". This was how Fiat trained its future executives at the time, "initiating" them in this way.

This image [Fig.7] takes us forwards in time, to 1977. It is a photo of a very important conference on colour held in Milan for the *Colordinamo* group. The man sitting next to me is Faber Birren, the greatest populariser of the modern culture of colour, whom I invited to Milan a couple of times and with whom I stayed in contact for many years. Birren founded a famous library at Yale University, wrote over fifty books on the subject and promoted knowledge of colour and its scientific and industrial applications, documenting its entire history. One day, I realised that one of his books, written about a year after our first encounter, featured the photo of a young couple dancing on a cube during the 1967 opening of the Electric Circus, New York's first big discotheque. It was my first trip to New York and I was with a young woman wearing the first nude look vintage dress ever seen. There was a chance of some millions to one that I would find myself on one of his books. He had chosen that colour photograph thanks to my psychedelic jacket that I had bought on the King's Road in London: a striking suprematist colour scheme, half dark blue and half lizard green. Here we have another legendary photo: it was taken during one of the two seminars on colour and materials that I organised. The entire Italian design world of the time was present: Achille Castiglioni, Bonetto, Busnelli, Zanotta, Mangiarotti, Bellini and many more. It was a very important event, as it marked the end of the idea that an object should be a single colour for its entire

Fig.7 Seminari Colordinamo, Faber Birren e i designer italiani al Museo della Scienza e della Tecnica, Milan, 1978. Courtesy Clino Castelli Archives.

lifecycle, establishing the thought that a product could be created and then chromatically changed ad hoc. A new vision of the design of colour, materials and finishes started to emerge. A few years later, the students at Domus Academy (especially the Americans whose design approach was "this is my project, now I choose the material") were shocked by the inversion of the paradigm, by the change of possibility. Today we are able to take inspiration from the material itself, as it is one of the languages of design culture. The relationship between shape, function and material maintains its value in terms of continuity within the project itself. Nonetheless, this may also be completely distorted by saying: "I will start from this material as it has in itself an immanent strength, an intrinsic character". We will look at this point in more detail as it represents one of the greatest changes in the history of design, a process that was also to involve the world of architecture, just ten years later. From this summary, we can highlight what I consider a new awareness of design, for which the "material substance" – defined as form, finish and colour – moves up a level. The form disappears and the colour, material and finishes become the true figurative reference points of the project: and CMF Design is born. By finishes, I also mean texture, which is another very strong factor of *No-form* characterisation, and which is to some extent still left unexpressed. Returning to the early 1970s, when we start to profile my professional history, what could a young designer who was not interested in designing objects do? He could merely hope to design meta-objects such as semi-finished products. In my case, the first semi-finished product was a material, and it was luminous to boot. Abet Print's *Lumiphos* is still listed in their catalogue after more than 40 years. It is not their greatest seller, but it remains on the market as it is a product that stands beyond any temporal logic of consumption. Here we can see the layered "sandwich" aspect of this semi-finished product, and how it could also be used in a performative sense. In fact, the image shows a surprising "reverberating bedroom" completely covered in *Lumiphos 14-580* (including the floors and ceilings). The *Glowing Chair* [Fig.8] was a sort of ready-made composed of an El Lissitzky chair whose seat and back were covered in luminescent laminate, obtained from a mould for outdoor chairs. The two components have simply been put one on top of the other, like an *accrochage*, thereby without any coherent formal research. The *Glowing Chair* has never been duplicated, but has appeared in a number of design magazines thanks to its iconic impact. Here we are in 1985-86. I had used Enzo Mari's 1972 chair – *Box* – as a reference point, since for me it was the most attractive *No-form* chair created by a designer, an object in which there is no compositional research. After about a decade, while working for Castilia who manufactured this object, I asked Mari's permission to make three prototypes in luminescent plastic, obviously presenting it as a tribute to his original project.

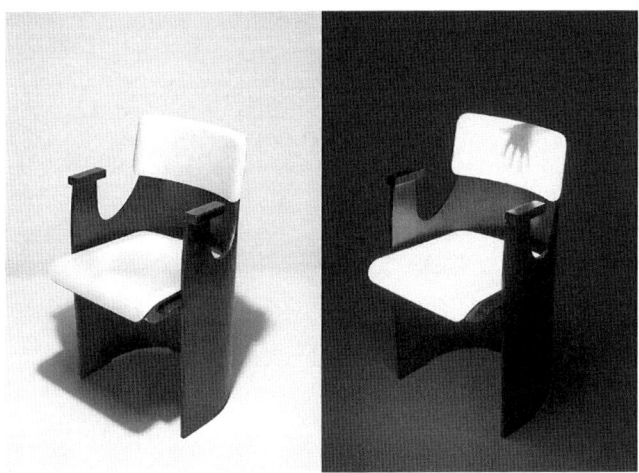

Fig.8 Clino T. Castelli, Glowing Chair, Lumiphos 14-580 Abet Print, 1972.
Courtesy Clino Castelli Archives.

Returning to the early 1970s, we see a piece for an exhibition that was to be produced by Abet Print on the theme of plastic laminate as a "Neutral Surface" [Fig.9]. The term *neutral* on its own leads us to the dimension of a weak formal characterisation, one that is tranquil and quiet, alluding to the decorative nature of these "paper-form" materials. I had written a text for this 1972 exhibition entitled "*Antifungus*: Technical logic of reactive surfaces" which alluded to "touch" objects, products that were by now completely without mechanical movements. I sustained that one day, when television sets would have become two-dimensional – completely flat – we would have entered

Fig.9 Clino Castelli, "Superfici Reattive", in L'invenzione della superficie neutra, Elementi Abet Print, 1973. Courtesy Clino Castelli Archives. Courtesy Clino Castelli Archives.

the era of "reactive surfaces". Today, for example, new O-LED technology introduces us to this extraordinary world that I had so accurately predicted, where not only does the form disappear, but also the object itself.

I know I promised myself I would limit this story to projects done before the 1990s, but here I have included one that I started in 2008. It is a wireless switch [Fig.10] that was successfully mass-produced by a world leader of home and building automation. It is an object of design in the solid state, wireless and without mechanisms, like the

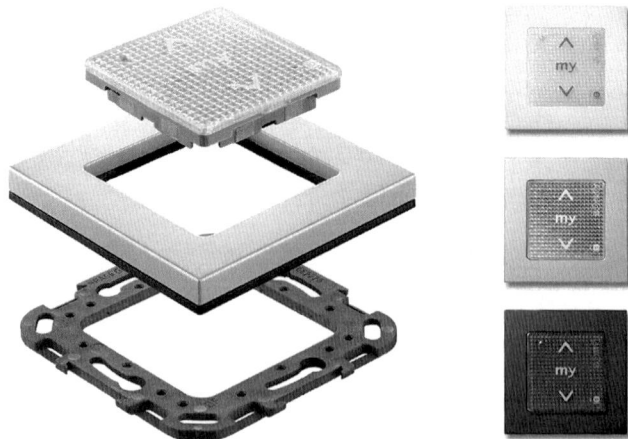

Fig.10 Castelli Design, *Smoove*, Somfy, 2010. Courtesy Clino Castelli Archives.

one I described in *Antifungus* and patented forty years after. As you can see, the large button is a reactive surface of capacitive type, sensitive to touch and remote controlled via radio. Dematerialisation has always been a great dream, and not only of technologists. The loss of materiality is increasingly seen as a "smart" evolution of the world of objects. At the same time, however, everyone would like to re-materialise things, for the intrinsic satisfaction of touching, of interacting in an analogical way. Nonetheless, when we are talking about whether or not something is needed, the fact of whether it is material or immaterial becomes an extremely important criterion.

The *Sistema Meraklon Fibermatching* 25 was developed in 1975 with Massimo Morozzi and Andrea Branzi, a project that is famous for having been awarded the Compasso d'Oro and which concerns the *Montefibre Fibermatching 25* [Fig.11] system applied to new synthetic fibres. In the process, they used "colour paste" in the material production phase, resulting in a limited range of primary colours which – when mixed appropriately – led to the desired chromatic results. I had created a partitive colouring method, made up of brightly coloured fibres mixed with other fibres in a special scale of neutral grey: a very ecological and structured approach that led to the formation of the colour only on a perceptive level, without the physical blending of pigments in

Fig.11 Centro Design Montefibre, *Sistema Meraklon Fibermatching* 25, 1975.
Courtesy Clino Castelli Archives.

the polypropylene filaments. This meant that no environmentally damaging colours or dyes were used. A decade later in Japan, I applied this method to the nylon fibres used in flooring, achieving even greater success.

During the same period, and once again at Montefibre, the *Colordinamo* series of manuals [Fig.12] was published, starting with the languages and historical trends that would involve the industries and the world of design, thanks to a new culture of colour that differed from the transcendent one of the world of fashion and style. When we talk about colour and materials, in many ways we use the two terms synonymously, even if the material – at the moment of production – is often only characterised through the identity of its colour. These were incredibly exciting years, full of innovations and initiatives that would contribute to the confirmation of Italian design's place in the world. One of the few huge failures of my career, but which initially was even more so for Dupont, happened in the late 1970s and early 1980s, when Kevlar was created by Dupont and developed and applied to replace steel in the construction of tyres. This material, in fact, while having the same strength, was seven times lighter than steel. Millions of dollars had been invested, and when everything was ready for production to begin, they discovered that Kevlar doesn't stick to rubber. There was no way around it, and in fact it was an economic disaster. A number of other applications were attempted, almost all of them in the military world. At the time, Louis Vuitton was my first inter-

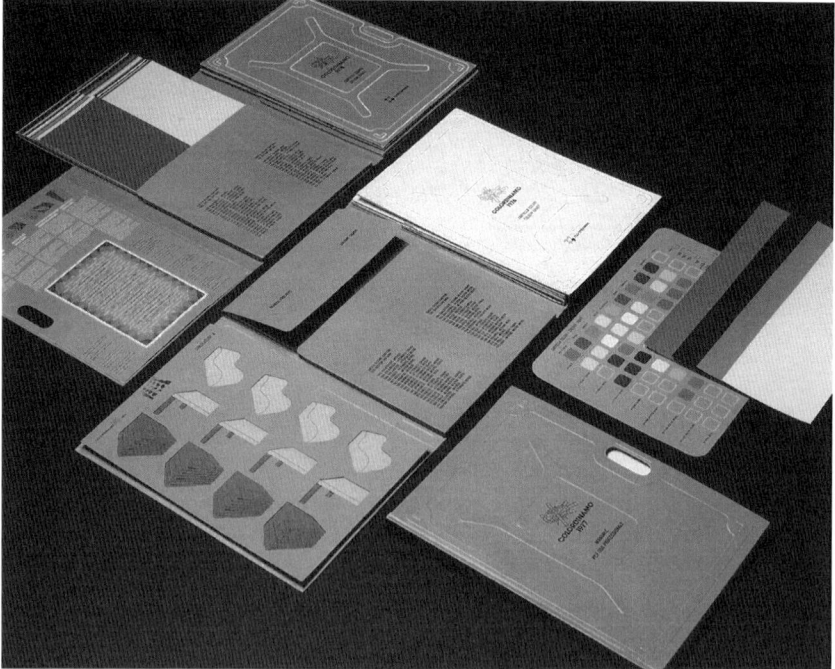

Fig.12 Centro Design Montefibre, *Colordinamo* manuals, 1975-77. Courtesy Clino Castelli Archives.

national client and we were studying bags and suitcases for them. I proposed the use of Kevlar, which to me seemed to be the perfect material – and almost impossible to copy. But when we created a hard suitcase [Fig.13] covered in this material, and tested it in continuous trips between Paris and Frankfurt, we discovered that the extremely strong Kevlar was in fact not very resilient at all. After fifty trips the material became as fragile as a normal piece of fibreglass. This is a perfect example of the meaning of innovation: it requires unbelievable effort, in which there is also a huge element of luck. In fact, despite the clear debacle, the production process we invented for Vuitton in canvas was widely used in the nautical field, precisely in the moment when Louis Vuitton was becoming the legendary sponsor of the America's Cup – two factors that put the firm in the limelight.

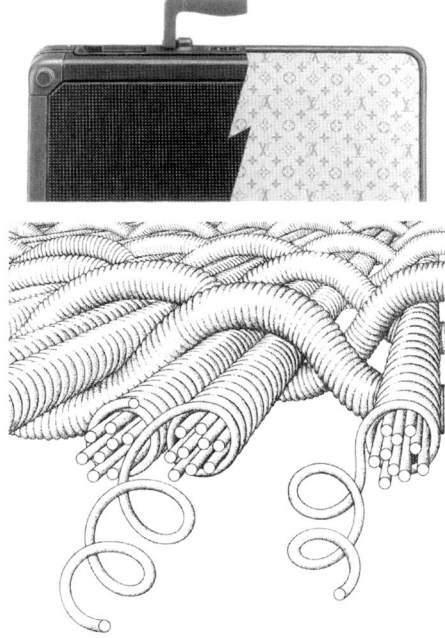

Fig.13 Clino T. Castelli, *Nouvelle Toile*, Louis Vuitton, 1979. Courtesy Clino Castelli Archives.

The "Kevlar case" did not compromise my relationship with the Maison; in fact, Louis Vuitton recently asked me to design an experimental object for their "Objets Nomades" programme. I proposed the *Lovebench* [Fig.14], a foldable bench that dips in the middle, for "amorous couples". This is another example of a typical *No-form* object: simple but not minimalist, considering the complexity of the puzzle it involves. This project also took an ironic look at the latest trends of additive technologies, as it is made up of dozens of layers of natural *Nomade* leather and shiny brass, overlapped and hinged to make the seat compact. These two materials, at the very heart of the Vuitton identity, are emotionally charged as they represent an iconic concentrate of *gravitas* in luxury, and are influential and classy.

Fig.14 Clino T. Castelli, Lovebench, Louis Vuitton, 2012. Courtesy Clino Castelli Archives.

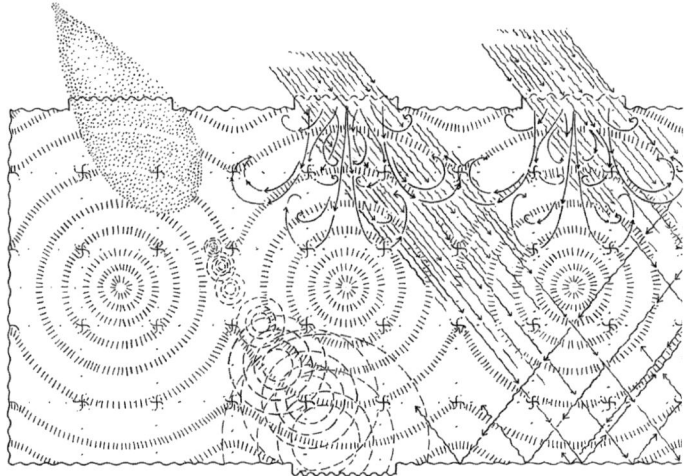

Fig.15 Clino T. Castelli, *Gretl's soft diagram*, 1977. Courtesy Clino Castelli Archives.

Gretl's soft diagram from 1977 [Fig.15] is of great importance for me. It was created during a kind of flash of inspiration, after having seen some images of the house Ludwig Wittgenstein designed for his sister Gretl. The images were in a book by an Austrian architect, who after learning of the impending demolition of the Viennese house, rushed to make an accurate photographic survey. In his book I discovered that Wittgenstein, with the help of another architect, tended to focus the project more on the effects of environmental qualities that pervaded the spaces than on shapes and composition. In that building I found something that went far beyond the neoplastic characteristics of minimalism, but which wasn't a part of the classic rules of Viennese architectural culture of the time. It was an *objet filosofique*, which expressed intentions and thoughts that were to be researched in the most immaterial factors, that I thereby wanted to represent. It was enough for me to establish where the natural and artificial light came from, where the heating and ventilation systems were, if there was a cooling system, etc. I can describe the acoustics perfectly without even setting foot in that environment, because having gathered the formal and material data of the space (photos and surveys), I know what the sound effects, the reflections and reverberations will be. All of this I then translated in figurative and poetic terms. The house was for Wittgenstein's sister, a beautiful and independent woman, a personality in Vienna Secession-era; she was even painted by Klimt. This diagram also tells her story, like the heady aroma of the perfume that lingered in the reception room, emanating from her boudoir doors.

In *Gretl's Soft Diagram* even the colour changes are represented by different sign of wavelengths. I consider this 1977 work my manifesto of Design Primario, because it deals with very weak environmental conditions which, if distributed on a broad scale, can have a strong effect, even of plastic nature. This may be the case of a single colour, which if multiplied over and over takes on extraordinary impact. One of my experiences was linked to the Japanese Mitsubishi lorries, whose colour was changed every few years. When they asked me for a new proposal – always regarding non-colours, in my case a *greige* – the event took on such importance that I had to return from Milan to the Kawasaki factory specially for a single hour of final prototype approval. That colour alone, in fact, over the years gave the brand an indelible identity on Japanese roads. The *Colorterminal* [Fig.16] is the centre of Creative Colorimetry where I developed the concept of colour as a design culture. Though I am referring to colour, we were actually already using the acronym CMF, which emerged from an important project

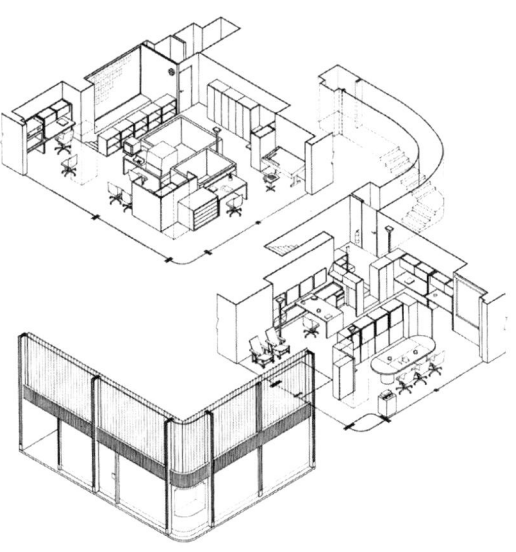

Fig.16 *Colorterminal*, 1978. Courtesy Clino Castelli Archives.

for Herman Miller USA. In 1980, two of their vice-chairmen came to Milan with Rolf Fehlbaum from Vitra, with whom I was working at the time. It could have been a very difficult meeting, because I had just participated in a competition, invited by their English branch to create a more European look for their *Action Office* system. I lost the competition because I proposed something that, for me, had to be done, but which in Europe couldn't be done: changing the base colour of the entire system, a greeny-beige typical of the 1970s. The American owners at that point realised that this was exactly what they needed in the USA. So I left for Michigan and I worked there for seven years on large office systems. It was a meta-design project based entirely on materials, colours and chromatic schemes, and it revolutionised the sector. In fact, from 1981 to 1985, they engaged in the biggest soft re-design effort of their industrial history. This involved a number of basic colours in dozens of different materials, and it was made possible only by massive use of computers and data processing systems. Today in Herman Miller there is still a department of CMF Design

The "Statement on Emotional Experience" [Fig.17] is a double-page advertisement for Herman Miller which appeared in 1983 in the New York architecture magazine *Blueprint*, at the time of the launch of the *Action Office* system re-design. My manifesto set out to demonstrate *No-form* years before its actual definition, where I theorised a design based on factors other than shape, such as light, sound, texture and temperature. "The

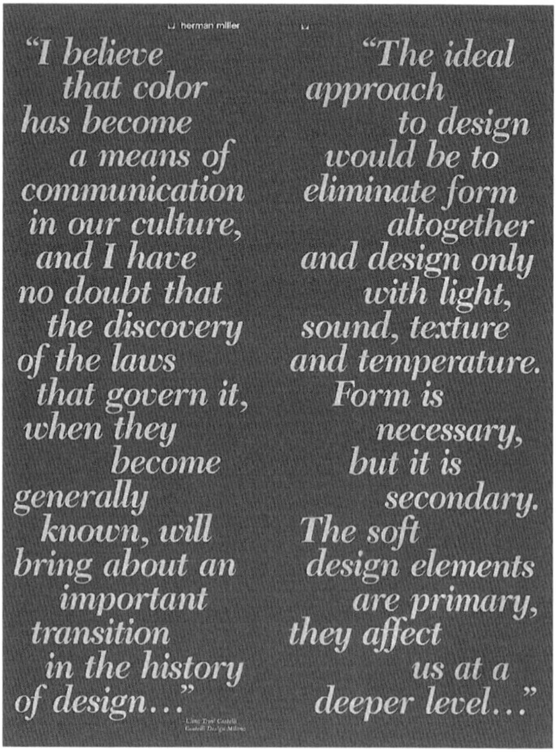

Fig.17 Clino T. Castelli, "Statement on Emotional Experience", Blueprint Magazine, October, 1983. Courtesy Clino Castelli Archives.

"I believe that color has become a means of communication in our culture, and I have no doubt that the discovery of the laws that govern it, when they become generally known, will bring about an important transition in the history of design..."

"The ideal approach to design would be to eliminate form altogether and design only with light, sound, texture and temperature. Form is necessary, but it is secondary. The soft design elements are primary, they affect us at a deeper level..."

ideal approach to design would be to eliminate form altogether and design only with light, sound, texture and temperature. Form is necessary, but it is secondary. The soft design elements are primary, they affect us at a deeper level [...]." This vision aimed to stress what I had sensed ten years ago with the *reactive surfaces*, describing the dematerialised destinies of future plastic and figurative languages.

This photograph [Fig.18] was taken under the light of a 150 watt bulb, which alone lit the entire dome of the new showroom I had designed for Cassina. That space, seen through today's eyes, is a precise expression of *No-form*, evident in the great catadioptric reflector of the cupola. Here, the colours of the reflections are virtual, just like the three-dimensional form of the lacunars. The curved wooden support panels are real, attached to the black background of the cupola. The different direction of the catadioptric 3M films creates different forms of reflection, depending on the impact of the light, generating extraordinary effects. This material is 10,000 times more reflective than white, meaning that if we were to photograph normal white and compare the two, the normal white would appear almost black. This also starts from another material: the triangular reflectors in prismatic methacrylate used on lorries, which I had placed on a wall at the Turin Automobile Show in 1965. In this case, the material is a retro-reflective textile used in the thin strips seen on the edges of New York firefighters' jackets. I ordered large rolls and I used them to cover enormous surfaces, obtaining the effects of "grey light", an interior reflection that was unknown until then. Here we can see how important the vast scale of the environmental phenomena involved was for *Design Primario*.

Fig.18 Clino T. Castelli, *Grey Light Pavillion,* Cassina showroom, 1985. Courtesy Clino Castelli Archives.

Perhaps it is no coincidence that just three years after the opening of the Qualistics Centre, Fiat became the leading European manufacturer in its market segment. All the development, inspections and controls on textures, harmonious compatibility between different materials, etc., were done in that laboratory. Those years were in fact very important for the research on new forms of quality that Fiat applied to great advantage. In Turin, I was given carte blanche and I created this Centre [Fig.19] which became an extraordinary design tool for the tactile, chromatic and osmic aspects of the materials, plastic materials in particular. I have always considered creating research and development tools to be a very important aspect of how I work. This approach affected the entire structure of the Qualistics Centre: even the ceiling inside the laboratory was made up of a giant catadioptric system that retro-reflected a very special grey light. Still today, in the Fiat Research Centre, there is a section dedicated to qualistics, or rather all those parameters of the perception of quality that lend themselves to typically subjective evaluations.

In the second half of the 1980s, a second ecological movement began, whose natural area of development was in materials themselves. Beyond the natural materials of the wave of environmentalism of the Seventies, we were now discovering artificial materials, man-made products that unlike synthetic materials came from natural substances. In particular, I had worked a lot on reconstituted wood. The first such products were made with African woods – quick growing but soft – which were recoloured and recomposed to obtain new decorative surfaces with both natural and artificially "syntactic" appearance. Here we can see Alpi's wooden flooring, which at the time was a great novelty as it was harder than the natural wood from which it was made [Fig.20]. An example of *No-form* design in which there is clearly a precise anti-compositional position is a furniture series I proposed to Cappellini for one of my installations in

Fig.19 Clino T. Castelli, *Centro di Qualistica,* Fiat-Comind, 1985. Courtesy Clino Castelli Archives.

Parq. Anche per gli occhi, non solo per i piedi.

Fig.20 Clino T. Castelli, *Parq Flooring*, Alpi Limonta, 1988. Courtesy Clino Castelli Archives.

"The Garden of Things", an exhibition during the 18th Milan Triennale in 1992 [Fig.21]. I designed some simple boxes, which appeared devoid of formal research but were rich in texture thanks to the recomposed woods designed for Alpi. My piece confirmed the importance of moving on from the design of shape to that of material, in which I spent a lot of time looking for the correct type of wood: the *Backing* clothes rack is a simple box without doors, which rotates against the wall in a very limited space; the *Tectonica*, an earthquake-proof bookcase created for Japan, inclined conversely so that no books fall in the event of small tremors. This bookcase shows the relief *Tectonica*, a brand new texture design that I discovered had never been created before. These *No-form* objects are often accompanied by a "puzzle" component, as in the case of the *Semestri* chest of drawers. This was also a non-minimalist product, in which the drawers, in a "distressed" wood that I called *black loft*, are hidden objects that disappear under each other in a surprising way.

These are just some of the many factors that have gone into how I think about and create design, from the very beginning. Certainly for me the most important factor was finding confirmation of my original hypothesis that materials would become immanent, assigning importance to the emotional value ingrained in the product itself. I believe that if we want to understand the meaning of *Mater Materia*, we really should look at matter as the mother of all forms. If matter expresses itself so well in the rarefaction of form, this shifts the accent to the expressive power of surfaces. And this directly links the *No-form* idea to the culture of design, that great vision that we have cultivated so well in this university. It is an asset that must be safeguarded and developed, beyond the on-going mutations of the major contemporary figurative languages.

[The text is an extract of the lecture "Mater Materia" by Clino Trini Castelli, 11th November 2014, Aula Castiglioni, Politecnico di Milano]

Fig.21 Clino T. Castelli, *Megatexture*, Cappellini, 1992.
Courtesy Clino Castelli Archives.

Fig.22 Clino T. Castelli, VSP - Cloud-computing platform, Hitachi, 2011.
Courtesy Clino Castelli Archives.

ATOMIC DESIGN AND THE ARTIFACTUAL ELEGANCE

ROBERTO CINGOLANI
Istituto Italiano di Tecnologia, Genoa, Italy

The *atomic design* approach used in the IIT is based on the "project to copy Nature". In fact, by "copying" we are doing the exact opposite of what designers do. But by copying the most evolved system in the universe, the act of copying becomes a challenge. Today, our message is *the scale of complexity*.
Over 3.5 billion years, Nature's evolutionary system has given it a perfectly functional architecture and has established a unique relationship between shape, complexity and function, which is what allows us to exist.
It has also created "objects" that consist in thousands of atoms, which we call antibodies: "hospitals" that circulate throughout our body and when they find a sick cell, they attack it biochemically, diagnose it and try to repair it. Over millions of years,

Fig. 1 iCub, the humanoid robot developed at IIT as part of the EU project RobotCub.

antibodies have been chosen and developed to be effectively selective. However, we have been able to show that this system is not yet completely perfect. In fact, where there is a mutation, or should something different occur within the organism, it often means that the antibody is not ready or has not evolved quickly enough, leading to a problem in dealing with the mutation. The problem of evolution lies in its slowness.
We lament genetically modified organisms as we think that they are dangerous, but in fact all of nature is a genetically modified organism. The difference is that when we modify organisms, we do so in a very short time - two to three years - while nature takes millions of years. Therefore, Nature's weakness lies in its slowness. And as it takes a long time for modifications, so also does it in adapting itself; while technology evolves quickly and genetic modification is too fast.
Increasing the scale of the complexity, and therefore the size, these "objects" [Fig.1] become bigger, millionths of a metre, and millions and billions of atoms give shape to eyelashes, wings, paws. From a purely biochemical interaction we reach a biomechanical cycle, because these objects, thanks to their complexity, start to produce a mechanical work; they move, they produce Newtons of force and, obviously, their size and complexity move towards a macroscopic regime: we are speaking of powers of 10^{18}, 10^{20} atoms. We move from small and simple systems towards more complex systems in which biochemistry and biomechanics are two sides of the same coin.
Be careful to not underestimate one of the most advanced objects that exists on the planet: the plant, an object that is anything but immobile. The tree is a great immovable "contraption" that, however, has an underground biomechanical intelligence; its roots find suitably hard areas with the correct pH balance, humidity, temperature and salts. Abilities such as gravitropism, thermotropism, and hydrotropism characterise this organism, which may be considered one of the most intelligent organisms created by nature.
Gradually, as complexity increases, we arrive at mankind which has a higher level of complexity: cognitive complexity.
In this observation of Nature, we have passed through three large disciplinary groups to reach this increase in complexity: biochemistry (chemical sciences), biomechanics and cognition that develop together to form human knowledge; not separated but unique as it is in nature, and as is the approach that we use.
The first responsibility of those who copy Nature is knowing how to copy its architecture, which means observing it and copying it. I can assure you that understanding architecture of five, seven or ten atoms is not easy, and it is not easy to reproduce it and re-establish that unequivocal connection between body and mind, which is what makes the difference between an actuator and a computer when they are not connected. The indissoluble body-mind connection is something that we are able to express through the concept of *morphological computation*: the mathematics of morphology. Our body is created to have a system of actuators that allows our intelligence to develop in a certain way. If we know how to do certain things it is because our intelligence has learnt how to make the most of our morphological and muscular abilities.
How can we copy this connection between mind and body? How can we produce an artificial system? To answer, I will give you some pearls of wisdom.

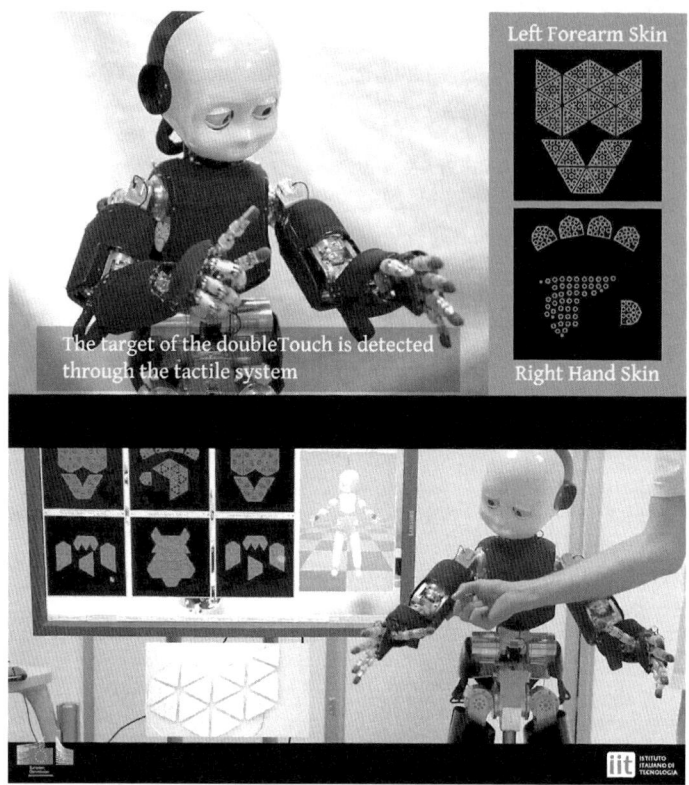

Fig. 2 iCub and view of the network of capacitive sensors in the iCub's hands and foreharms skin.

Fifty years ago, a "guy" was studying the resistance of germanium and created a new object, the "resistance conveyor" which allows electric current to pass from one side to another, following a direction. This new object, the 2.5 cm transistor, at the beginning was useless for all purposes but winning the Nobel Prize. Today, any chip contains around 150 million transistors. As each of these objects is an on/off switch, it may be considered the qualifier of a code, of a language that processes information. With a million transistors that go from 1 to 0 in extremely quick frequency, I can simulate a brain that processes billions of pieces of data per second. Therefore, the first step is the understanding of the computational unit.

Today, the computational power of machines has reached a billion billion operations per second. But in order to obtain computational capacity that can computer simulate a machine, an airplane, a crash test or to produce medical images, we need 20 megawatts of electrical power – approximately that consumed by a city. On the other hand, our brain makes a billion billion binary operations per second and consumes just 30 watts: there is a huge different in the architecture.

Our brain is made up of carbon, water and neurons. It is self-learning, it teaches, adapts and consumes very little. With just a square of chocolate it can continue working for two days. Every neuron contained within communicates with another 20,000 neurons in a three-dimensional box.

On the other hand, a super computer uses silicon, functions with copper wires, is flat, very fast and very stupid as it is merely repetitive and requires megawatts of energy because the transistors can only communicate if in close proximity. If a transistor is very far away, the electrical resistance requires more current, and therefore a large object requires an enormous amount of electricity to work.

The first thing to copy from nature is its computational method. Currently, our knowledge goes from the brain (carbon and water) towards silicon, against which we lose energetically speaking.

Speaking of *iCub* [Fig. 1], the robot created by IIT's researchers, which was given a face based on human facial expressions, is able to carry out around 2-3 million operations per second. In terms of computational power, this quantity is very low. A super computer, iCub could reach the same operational quantity as humans, but it would require its own personal electrical power station. To give you an idea of the dimensions: to create an artificial mouse would require the capacity of a million billion operations per second, the equivalent of a very high-performance machine.

So in IIT we thought up a different strategy: the intermediation of an *icloud*, where all the intelligence of these machines is kept. With a very fast wireless transmission, machines can quickly access this "global-cloud". We have thus resolved the problem of computational capability.

The second pearl regarding the connection between body and mind.

The first thing we asked from a designed machine was that it was social. Machines must be able to interact with man. If I touch and push someone, they will move away from me. If I attract them towards me, they will approach. We have tried to understand from the physical point of view what type of interaction rules these mechanisms. In order to interact, we need sight and skin.

The iCub skin is a neoprene with a network of capacitive sensors: around 12 sensors every two square centimetres are required to be able to feel not only the presence of a person by touch, but even if we act with a force of pressure or traction (with vectorial indication), recognising the direction of the force [Fig. 2]. This system is what makes the machine social. The machine is able to move itself and accompany the movement. It is dynamic because it has a body with 53 degrees of liberty - approximately the same as our own. It is equipped with skin like man, and uses the neuromorphic model of vision. This model is what regulates our vision. For example, right now I am speaking and moving a small part of my body: my head and arm. The rest of my body is still. Have you ever asked what happens in moments like these? How does vision act?

Man does not accumulate the frame of the image in front of him in every separate instant if it doesn't change as otherwise, having to sum up this image every instant, his memory would finish very quickly. What actually happens is that man adds to his memory only what changes, he carries out a differential reading. He only remembers what has changed. In this case: the hands, head and mouth. Everything else is held in the background until something moves, and as it changes it is memorised.

Making a machine work like a human being means copying this function: giving the machine a neuromorphic system of vision and touch and imitating those strategies that the human body has saved throughout its long evolution, in order to become quick and adaptive with the lowest energy cost possible.
This also is our task in IIT: studying the brain to understand how it works, translating this information into a neuromorphic computer system which will be useful in making our machines work. It is a study that goes beyond electronics, engineering, biology: it is a global science.
Thus operating, our machine evolves and learns with us. iCub has spent a number of months learning to write. It learns like we do: it sees a letter, sees the movement of the hand and then tries and tries again until it perfects this movement and in the end it learns, just as a school-age child learns to write.
iCub writes its own name - with the same uncertainty typical of children - because this is what happens when we can feel a pen between our fingers, when we have a stereo vision that gives us a perception of three-dimensional space, knowing how to regulate force to be able to write with the pen.
This is the first step towards connecting man and machine, body and mind.
Obviously, the simplicity of this manufactured product, the fact that its function and its shape had been studied to do certain things, allows us to develop an increasingly sophisticated intelligence. This is the relationship between artificial intelligence and intelligence of the material, of the design of the form and function. We can start to challenge the child by asking him to do something that no one has ever asked him to do before; for example by giving him two Lego blocks and letting him find how to put them together, without any control strategy, without anyone telling him what to do. In this exercise, iCub found the most economic strategy: in 7-8 minutes he adopted the strategy of putting the two corners together and turning his wrists.
At this point we cannot stop copying. Nature has given us a long series of lessons that we have to put into act.
Man has a vestibular system that allows him to resist very intense solicitations and to vary his set up, to keep his balance. In order to obtain a similar pneumatic system we need to copy the movement of muscle - Achilles' tendon - the elastic adaptability managed by the vestibular system of the ear. This is a synergic muscular system with a neuromorphic view, a sort of oleo-pneumatic system that allows the machine to move in completely random situations. In this example, the robot does not have the sense of sight [Fig.], his is pure self-balancing locomotion. Just like a blind person who walks and instinctively stays upright thanks to his vestibular system. A huge engineering project whose movements are an analogy for what occurs in nature. A fascinating project. We would like it if these objects could help human beings throughout their daily existence.
We also designed a "plantoid": the first robot plant on the planet. It has the intelligence of a plant, leaves that move towards humidity and light, roots that travel downwards like normal plants (gravitropism), thanks to an incorporated gravity sensor. Other very important characteristics are tigmotropism, when plants move in response to tactile or contact stimuli, so that the root feels when it hits something and it is able to avoid involuntary impact; hydrotropism, or rather the ability to sense water; and thermotropism or rather the ability to move away from heat sources.

Of nature, we also imitate the growth mechanisms that from above move downwards up to dozens of metres.
What is this study useful for? We should we copy its roots?
A possible application in the biomedical field is the "endoscope of the future": a tool that can find tumoral pH receptors, or identify what is needed, leading the point of the root where it can only go surgically. The basic idea is that the sensorial system of roots goes where we want it to: an unreachable recess - following biochemical mechanisms - in order to take the laser, chemotherapy, anti-inflammatory and anything else required for treatment to the point in which it is needed.
Another possible application for the plantoid is aerospatial travel to other planets: once we have arrived on an unknown surface, it is possible to activate a feeding system and at this point strike root and study the terrain.
Nature has given life to a multitude of solutions of which we ourselves are unaware but which are already ready. We just need to copy them.
In this fantastic relationship between body and design, natural life and copy, we also take inspiration from very small elements such as atoms and molecules.
By doing this, we have developed a nano-tech sea sponge, which like the natural ones has capillary ducts that absorb water and impregnate. We then changed the chemistry of the sponge with some small objects, a few atoms, to obtain an object with some impressive characteristics: a sponge that repels water (super hydrophobic behaviour) but absorbs oil. When removed from the water from which it has absorbed the oil, the sponge may be squeezed to remove the oil and then returned to the water. We call it the "nano-tech mussel" as it behaves like mussels in the Taranto Sea.

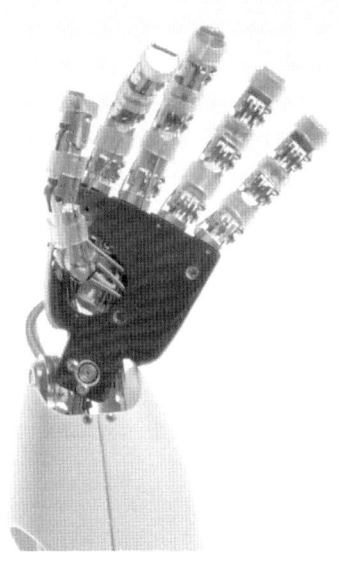

Fig. 3 iCub developed at IIT.

Through the same method, we can determine the sponge's characteristics should we want to go against evolution, step beyond the copy, and be even more "challenging". We can decide that the sponges absorb heavy metals - we just need to decide the biochemical interaction that you wish to activate or deactivate.

The problem of reusing water is very important and these are solutions that come from evolutionistic trajectories. Just think of how important it could be to use this technology in places where there is a lack of drinking water. Humankind consumes 2,600 cubic kilometres of water every year and in fifteen years this will rise to 4,000 when the population reaches nine billion, and water tables will reduce in quantity by 30%.

Finally, on the subject of copies, another serious ecological problem is the accumulation of plastic: 290 million tonnes of polymers produced by petroleum every year. These plastics biodegrade in around 1000-1200 years producing phenols and other environmentally-damaging substances.

So my question as a "copier" is: if we have to use plastic, can we not use something that works better and costs less than petroleum?

Observing trees, plants and leaves, it came to mind that everything made by nature is made of fibres, especially in the plant world which uses cellulose fibres. Cellulose is a sugar; in fact plants decay, dissolve and are not harmful. They transform into carbon, oxygen, hydrogen and nitrogen.

We asked ourselves how much plant waste the food industry produces globally: Europe produces 26 million tonnes, 10% of the global production of plastic. And we have found a new technology to recover cellulose fibre from plant waste and obtain plastic from it. In this way, we can produce plastic from parsley which biodegrades in three years, but also from potato, beetroot, spinach, etc. We can use plant plastic today to imitate PMS which is used in kitchen gloves, the PT of bottles, and cover all mechanical structures that we still produce today from petroleum, but with 100%-natural plant waste.

The final message that I want to give out is: we have reached the point at which we humbly understand that we have to copy what Nature has perfected: a series of processes and relationships between form and function which are so evolved and so advanced that only a madman could think of doing better.

Currently, here at the IIT, and just like the Japanese did in the 1950s, we are copying and hoping that in thirty or forty years our technology will be so reliable and so indestructible that we will then have even more to propose - beyond evolution. At the moment, it is too early to do so.

References

Roberto Cingolani (2014), *Il mondo è piccolo come un'arancia. Una discussione semplice sulle nanotecnologie.* Milan: Il Saggiatore.

THE GROWING LAB: FUNGAL FUTURES

MAURIZIO MONTALTI

Officina Corpuscoli, Amsterdam

For years I have been undertaking a design research that focusses on the production of material that takes its inspiration from natural life forms, their life cycle and the sustainability of the organic matter. Living organisms such as fungi are the main material used in my design. These microorganisms have always fascinated me. They are omnipresent: under our feet, everywhere we look, in the air we breathe. All the time we are inhaling millions of spores. But, despite the fact they are so present in human life, we have stalled in identifying them as elements of our life, just as we have with so many others that we associate with feelings of repulsion, disgust or danger, in some way diminishing the role that these microorganisms have in the ecosystem of which we ourselves are also a part. And their role is an important one. In fact, fungi are important within natural rhythms, like the cycles of decomposition of the various types of organic and inorganic substrata. They are the great recyclers of the natural world and thanks to their ability they cover a series of activities of transformation and undertake a variety of tasks that allow us and our planet to survive. [Fig.1]

Fig. 1 Schizophyllum commune growing on log Officina Corpuscoli. Courtesy Maurizio Montalti.

In the Italian language, "fungus" means both the living organism found in nature as well as the food stuff we serve for dinner and consume (mushroom in English – translator's note). This tends to create confusion. We must, therefore, make a distinction between the fungus-fruit and the plant organism that generates this fruit, the *mycete*. Also known as *mycelium*, the plant apparatus of the fungus is made up of a long network of filaments distributed throughout the soil which is responsible for "breaking up" the elements within the natural matter and restoring them in a different way to other life forms. This in fact is what mycelium does in nature. It presents itself under the form of long filaments, cells that are invisible to the naked eye which tend to grow lengthways in any type of degrading organic matter. The structure of mycelium is made up of individual filaments, known as *hypha*, a sort of fibrous formation similar to the roots of a tree and organised like a series of interconnected cells. It is also interesting to observe how they move, how they grow and how they connect. Some talk of mycelium as the first global network, which in many ways it is: a glue that binds the earth while at the same time transmitting information between the different life forms found in the soil. [Fig.2-3]

If you have ever walked in a wood, you will certainly have come across this type of manifestation, characterised by surfaces similar to cotton, quite spongy, which grow on trees and dead branches; this is merely the macroscopic manifestation of the combination of millions of *hypha*. On the microscopic level, an individual *hypha* is around 4-5 nanometres in diameter, giving you an idea of how difficult it is to see it. [Fig.2]

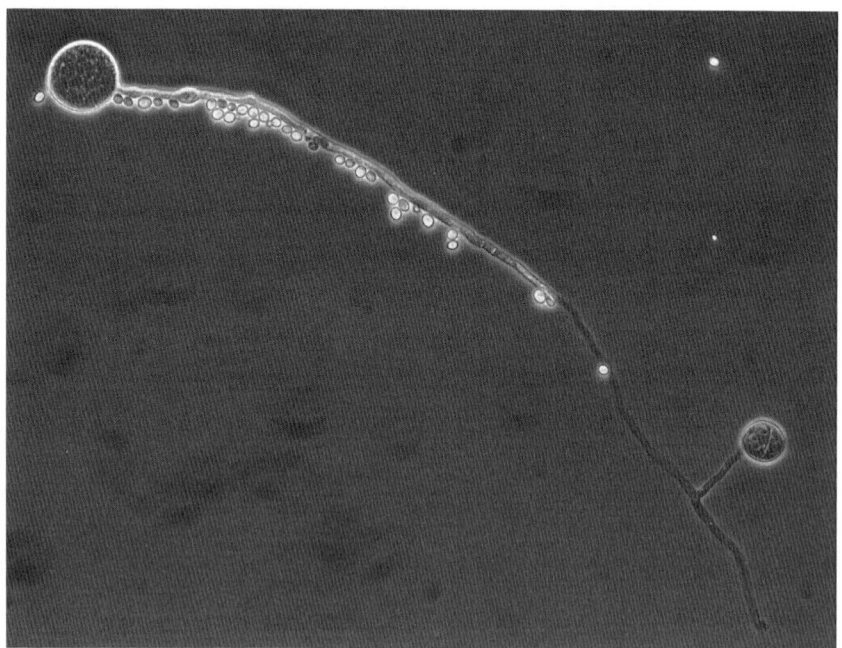

Fig.2 Single Hypha. Courtesy Officina Corpuscoli.

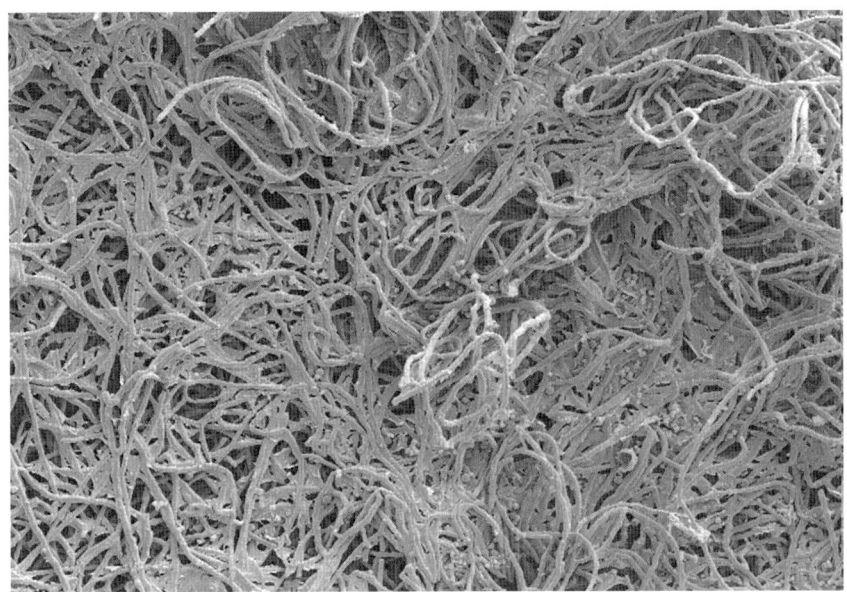

Fig. 3 Micro SEM mycelium network. Cortesy Officina Corpuscoli.

We still know very little about these microorganisms, but there are many indications that allow us to induce how many positive possibilities they may provide.

The reason that I have so interested myself in these lifeforms is the number of benefits that may derive from their use, if only we can overcome the prejudice that makes us see them as pathogens.

In the first phase of my exploration, in a sort of analysis between the philosophical and the social, my attention dwelled on the complementariness of the concepts of life and death: one cannot exist without the other in the sense that death triggers new life (Montalti, 2010). I was interested in the relationship between life and death and the analysis of decomposition cycles in the human body. For this reason, I tried to understand the role of these recycling organisms. I felt the need not only to gather evidence from the theoretical viewpoint through scientific publications and concepts belonging to matters of which I had no basic knowledge, such as biology, but being a design I also felt the need to get to know the matter in person. In the Netherlands, where I live and work, this is the main characteristic recognised in a designer: he is a maker; the designer is responsible for concluding the development of his projects working on things practically and directly, not only in the model phase (known as the mock-up phase), but also at the beginning of the production phase.

The need to deeply understand the matter I was dealing with led me to touch this matter with my hands, to better understand the behaviour and growth of these microorganisms. Mycelium therefore became my raw material, just like a traditional material, such as wood or metal, with the only difference being that I was dealing with a living matter which was therefore more complex insofar as it was difficult to predict its developments.

In my laboratory I started a series of experiments to understand the interaction between mycelium and matter. Initially, my exploration regarded the methods through which fungi would colonise certain organic materials such as cotton and other textile materials. This initial process of comprehension led me to create two projects that were the initial thesis works for the IM Master Course - Conceptual Design in Context of the Design Academy Eindhoven: "Bodies of change" [Fig. 4] and "The ephemeral icon" [Fig. 5].

As said above, one of these projects regarded the physical death of the human body seen as a substrata, as something tremendously ephemeral and temporary, crossing the emotional response that people have when they lose a loved one. If we concentrate on behaviour regarding death, the western model appears in all its euphemism that tends to deny death, as can be noted within the various practices of the funeral industry. Cremation, for example, frees us from the remains of the deceased with an enormous quantity of unnecessary energy consumption and then releases a huge amount of pollution and dioxins into the air we breathe. Burial on the other hand has its own methods: the deceased's body is often injected with a material aimed to preserve an appearance of life, then given make up, dressed and placed in a coffin with a series of other objects depending on the culture. These practices imply a reflection on the relationship between society, culture and nature that distances us from the prerequisite that humankind is part of the natural ecosystem.

I started asking myself why we tend to subtract ourselves from everything that characterises the natural systems, when on the other hand anything that dies in nature decomposes and is reabsorbed in terms of elements into the entire cycle of life. This is why I started working in the laboratory trying to understand how fungi may be the agents responsible for the degradation of human remains. I experimented on an epithelial fabric similar to human skin – pig's skin – and was able to continue a process of degradation but also of detoxification. Our bodies, in fact, due to the conditions that we have created in the society in which we live, are extremely toxic. It is a phenomenon that can easily be witnessed in other cultures, such as in India, where in some tribes the deceased's body is given to birds to eat. Today these birds die as they ingest toxic substances found in our bodies.

The interesting factor demonstrated by the experiment is that fungi are also detoxifying agents. The act like filters that purify toxic substances through their cellular walls, restoring cleaned substances. For this reason, they are also used in industry in processes to purify gaseous or liquid components.

Through this direct experience undertaken, and be talking with scientists with whom I worked in the laboratory, I tried to imagine an alternative to the traditional practice of burial that could allow us to think of death in a different way, leading to the development of a new funerary custom. This approach could seem macabre or even strange, but I think that designers must not only be trained in designing normal daily customs but also for critically analysing the social practices that we continue to perform, proposing alternatives to the needs or changes that we need in cultural terms.

The project that resulted from this reflection and my experiments, entitled *Bodies of Change*, consists in a shroud, a sort of cape that is placed over the deceased's body which helps to transfer the organic culture of a specially-chosen fungus to the exterior

of the body in order to facilitate a quick decomposition process and at the same time filter all the toxic substances that are found inside the body. I thus tried to explore the relationship between nature and the human body in a cyclic connection with the environment, because the function of the fungus is not only to purify the substances, but also to deconstruct the elements, separate and redistribute them as a source of new life, viewing the body as organic matter whose elements may be returned to the cycle.
Bodies of Change was born from a philosophical reflection that configures an alterna-

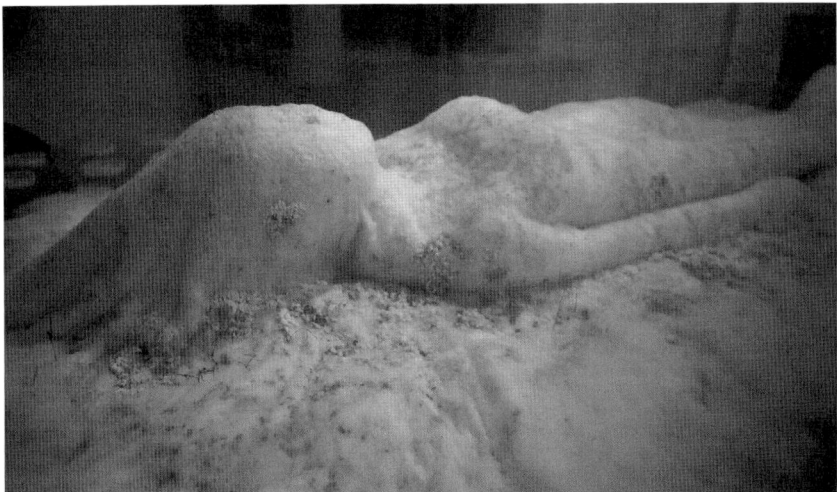

Fig.4 Bodies of change, project by Maurizio Montalti.

tive that is real but hard for the funeral industry to accept thanks to the current laws in force, despite the great interest shown by the industry itself. I had the chance to present this project in a number of contexts and the reaction of the public has been unexpectedly positive, especially the older public, who were fascinated by the possibility of being returned to the environment in a more unconventional way.
If initially my research focussed on organic matters, a part of my interest also touched inorganic materials such as plastic which, as we know, have a huge impact on the natural environmental as they are so hard to decompose.
That of plastics is a universe in itself. The 20th century materials, formal possibilities and economy have all brought us to the current state of "progress", while determining a negative impact on the environment of which today we are suffering the catastrophic consequences.
I tried to analyse the attitude of our "disposable" culture, and what happens when we use these materials every day, especially if we consider the enormous amount of packaging that we consume and throw away. Despite the fact that today there are recycling technologies and projects, plastic is often accumulated in tips in which there is no chance for degradation thanks to a lack of oxygenation (aerobic-anaerobic conditions). The materials are mixed, and once they are accumulated in a tip, the chemical elements

that make up these materials are redistributed; elements enter the soil and are often confused with food by other lifeforms thereby conditioning the entire ecosystem, all the way to us, as a part of what we eat. One of the results of this phenomenon is the pollution that lives inside us and which is part of the chemistry of our bodies.

Chris Jordan's photography project is particularly incisive in describing what I am trying to explain.

The British photographer undertook a project in the Midway Atoll, a small archipelago of islands in the middle of the Pacific Ocean 2,000 miles from the nearest continent. Here, Albatrosses feed their young the material they find in the infamous *Pacific Trash Gyre*, a continent of plastic material that rotates continually in the middle of the ocean. The plastic material is mistaken for food and obviously once the chicks eat it they die. It is a sort of macabre mirror of our times that leads us to reflect on one of the fundamental challenges we face today.

During my explorations in scientific literature, I came across a fungus that can degrade plastic, the *Phanerochaete chrysosporium*, widely used in industry as a filtering or purifying agent in a number of different processes. The subject fascinated me greatly, not least because this research seemed to have stopped despite its positive results. I humbly tried to undertake a small experiment on the subject. Thanks to a collaboration (which is still underway) with Utrecht University, I started working in the department of microbiology which gave me access to theoretical knowledge but above all to laboratory knowledge as well as the chance to use new tools which became my tools as a designer. I started carrying out a number of experiments in which the fungus attacked various materials, and I measured the degradation of the materials that I subjected to the action of various types of fungus (because it is of course a species, but different branches can be used and each one acts in a different way). One of these experiments, unexpectedly, brought some extremely positive results. In fact, a sample in which some rings of acrylic resin were inserted in a polycarbonate box was decomposed by the fungus. This is of particular interest as it is able to degrade some of the most resistant plastics such as polycarbonates, which are also particularly toxic.[Fig.5]

This was one of the most interesting results of the research, for the impact it has had on the construction of a view based on tangible scientific experimentation. I would like to make clear that the most appropriate definition is tangible, rather than scientific, in the sense that it was one of those experiments that was forgotten and rediscovered ten months later with the observed result.

We asked ourselves a number of questions: how did the material transform itself? What type of material resulted?

The analysis of some samples gave us the answer: they are basically made up of carbon and nitrogen; a kind of fertilizer. How did the rest disperse? What gases were created? It was impossible to know this due to a lack of constant observation, and this is why I feel it is important to stress that the scientific experiment should be repeated.

My projects aim to stimulate debate on the possibilities that the use of microorganisms may offer the society in which we live, and encourage other competent operators in science to continue this research; to encourage designers and the entire design field to think critically about the materials that are used every day.

In order to create a short circuit between the culture of 'disposable', the toxicity of plastics and the opportunity of plastic-degrading fungus, I decided to focus my attention on a globally recognised object: the "mono-block chair", the garden chair [Fig.6], an extremely cheap object that breaks easily (if you go to any tip you will find a huge amount of these broken chairs). This object is a declaration on the life cycles of consumer products, compared with the immortality of those materials of which the majority of consumer products are made. Once again, what I aimed to do in this phase of the project was highlight the importance of complementariness between life and death, the idea of a 'cycle'. In this project, I wanted to stress the idea of infusing

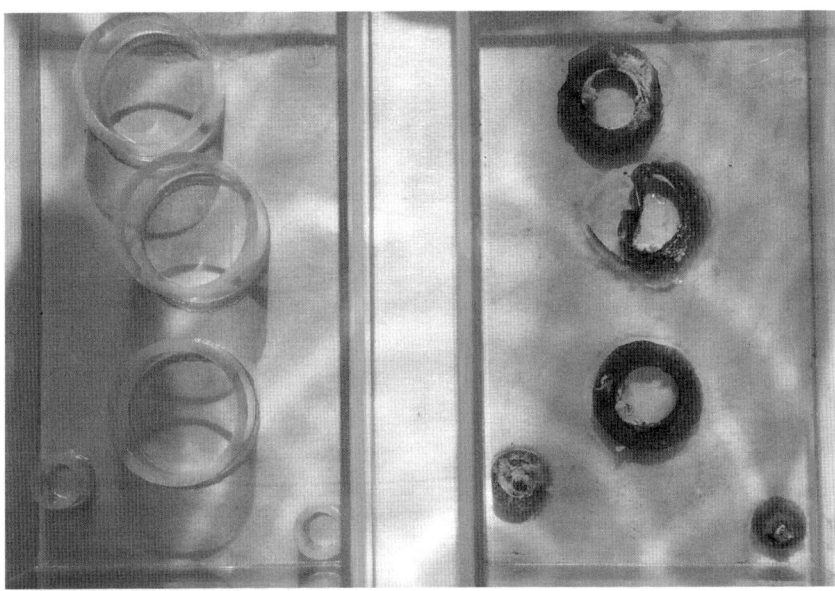

Fig.5 The Ephemeral icon-fungi-eating-plastic (comparison-start-end).
Courtesy Officina Corpuscoli.

life in a completely dead and immortal material in order to execute a process of total dissolution: bringing a "Frankenstein chair" to life so as to kill it because it is alive. This aim was consolidated in a project formally similar to *Bodies of Change* which instead of intervening on the organic material, does so on an inorganic and toxic one.

I therefore developed a shroud for the chair, a cover containing the culture of the fungus that can degrade plastic materials. This is made up of a sublayer of cotton, a hand-made felt, which contains further nutrition in order to favour the colonisation process of the fungus on the plastic and transfer the culture from the cover to the chair, until the fungus feeds on the chair. All of this may work in the laboratory as this type of experiment may be undertaken in stabilised conditions and a sterile environment in which there is no competition between microorganisms and the fungus can work in complete peace in constant and controlled conditions of humidity and temperature: in the real world this may not occur.

Once more, the goal was not to put myself forward as the person to solve the eternal problem of plastics and their impact on the environment, but more to encourage discussion, and thus introduce the public to the opportunities that may emerge if we "collaborate" with the microorganisms.

It is also in this way - more than in any other – that the role of the designer can express itself. He acts as a catalyst of thought, attracting the attention of an audience that is much vaster than the scientific community. The most interesting phase of this project was the success in taking scientific knowledge out of the laboratory. I wanted to develop, experiment and personally participate in these themes in order to bring them to a vaster audience in a simple, approachable way using clear contents, without necessarily using the scientific language that many do not understand.

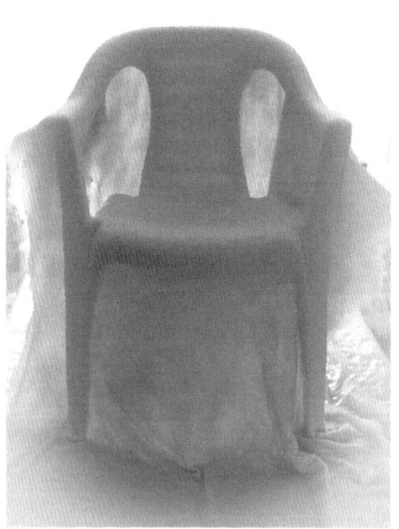 

Fig.6 Ephemeral Icon by Maurizio Montalti using the "mono-block chair" as icon-biocover-making process. Courtesy Officina Montalti.

The scientific field has become my new "workshop". Instead of working with plastics, wood and metal, I found myself working with completely new tools and different notions that I had to understand. An extremely interesting journey that I would recommend to all.

My research then continued with the *System Synthetics* project [Fig. 8], the result of a grant obtained in the Netherlands and through which I continued exploring the degradation of plastics.

The concept of this project was based on production methods and energy consumption. If we think back to how energy was produced during the first phase of the industrial revolution, we remember that large, complex architecture was required [Fig. 8] in order to exploit some natural resources. Today the methods have changed. In the field of research today we use microorganisms to produce different types of energy sources or substances, such as fungi, which are used as cell factories and for the development of consumer products. I was once again very interested in the possibility of intersecting the abilities of the plastic-degrading fungus with that of another microorganism: a yeast (*saccharomyces cerevisiae*) which has always been used in fermentation processes in beer production, wine production and in bread-making, and which is able to produce alcohol.

I thus tried to meld the abilities of these two microorganisms (*endo-symbiosis*) trying to create a sort of new life form derived from the forced, man-operated symbiosis of these two microorganisms. A new chimera, that can degrade plastics, decompose them and at the same time restore energy in liquid form – everything used to make the material (basically fossil fuels) which is transformed once again into a liquid energy form (bioethanol).

To this end, the designer comes to wear the coat of the scientist, to better understand the mechanisms of the fungal micro-world.

Basically, the objective of the project was to raise questions on the use of technology that is used today and which will be increasingly used in the future, or rather synthetic biology, on which everyone has radical opinions, both positive and negative; in any case, I believe (and this is something that I discovered myself) it is fundamental to talk about it. The project aimed to provoke. We proposed the creation of a new mutant lifeform created by man in the 21st century, creating symbiosis (a process that in nature takes millions of years) fusing a yeast cell with a fungus cell. Something that is very difficult to do in real life for a question of cellular synchronisation processes and size, even to be able to create this new lifeform to resolve a problem caused by humankind: therefore finding a solution created by man in order to resolve a problem created by man, asking the question if this type of method may be correct or not.

The aim of the project was to create collective understanding and knowledge in order to put a spotlight on a technology that already exists, that leads us on the continual classic trajectory of so-called progress, trying to then decide collectively which use we want to make of a technological tool that is at the very foundation of current research. This is the starting point of my reflection on how important it is that designers cultivate a passion for everything that makes up new technologies, specifically biotechnology. We are currently experiencing the start of a revolution. After the industrial and computer revolutions, the biotechnological revolution is when microorganisms and

Fig.7 System Synthetics project, installation overview. Courtesy Officina Corpuscoli.

life sciences will play an increasingly important role in the development of human "culture". I believe that designers have a great responsibility to understand everything that animates the development of these technologies and the manner in which they can or must not be used.

The final part of this project is made up of laboratory elements that were custom-made for my laboratory experiments. The installation aims to show the transformation from input – plastic – to the resulting bioethanol, through a fermentation process in which the plastic material is transformed into organic matter by the fungus that in turn ferments with the yeast that distilled transforms itself into alcohol (something that is possible in small quantities in a laboratory).

The resulting image that can be seen under the microscope is the visionary model of this new Minotaur.

Once again, however, this is not a problem-solving project: it does not aim to bring effective results or great discoveries, merely to encourage a process of competent research (a provocation to the scientific world on the need of exploring these processes of transformation) and above all to involve a – possibly vast – public through critical questions, to build up critical thought regarding the materials we use and the possibilities that may derive from the cultivation of microorganisms.

Based on these experiences, I later had the chance – mainly in Amsterdam thanks to institutional partners such as Mediamatic or Waag Society, which promoted the spread of techno-scientific knowledge through design skills – to organise workshops and various other activities in order to share the knowledge elaborated in the laboratory in a quick and more immediate way, with a community of designers, architects and artists interested in approaching this new paradigm of research.

The *Biologic* workshop was organised in 2013 with the Waag Society in collaboration with the Italian group Co-de-iT, a research collective on computational design, and with the Sonja Bäumel studio. During the workshop, we used generative mathematical algorithms to understand the behaviour of certain microorganisms and to try and guide the microorganisms towards the formation of different architecture. We built specially a 3D printer with WASP Project support (a small company that produces 3D printers that mainly use clay-based materials). It was interesting to be able to discuss the technological fantasy of 3D printing. This is, in my opinion, fantastic technology but it is also of no use to the consumer, despite all the talk there has been on the democratisation of production means, in which the consumer replaces the industry. There is much speculation on the matter, which I personally don't believe, as I don't believe that this forecasting will ever come about in such a clearly-defined way.

What disturbs me in particular, concerning the talk about this technology, is the concept of democratisation of the production method, which in the beginning was very interesting as based on open-sourcing, on the sharing of knowledge and collaboration

Fig.8 System Synthetics, a visionary symbiotic organism. Courtesy Officina Corpuscoli.

on a global level to develop a technology, but it is the materiality that characterises the problem. The majority of those who work with 3D printers, in fact, work mainly with plastics known as bio-plastics (which have nothing to do with anything biological), creating products which are the result of style exercises (an ugly style if I may say so) which lead to the production of great quantities of useless objects. Objects that do not even need to be produced with this type of technology, which uses up more energy compared to other processes for working plastic.

One of the main subjects to be discussed and to research is that of the materiality that distinguishes this technology: think of how we can use it using materials that may serve as a support, a structure, to allow the growth of microorganisms that will then replace the matter itself, representing nutrition for the organisms and leading to the formation of new organic architectures.

The *BioLogic* workshop allowed us to investigate a specific microorganism - slime mould - which has marked intellectual capabilities, and can optimise the distances between points. Slime mould is a fascinating organism. It is also used in research on artificial intelligence, where some models of robots produced in Japan are governed by the study of slime mould.

Research has analysed how this microorganism interprets the disposition of some nutriments distributed on a design generated by some computational algorithms that have in some way to do with the growth of organisms [Fig. 9]. The project is still underway in understanding the microorganisms' development models and how the microorganism reacts to the actions of scientists interpreting its behaviour. It is interesting to observe the results of the initial experiments [Fig. 9] in which we can identify the pattern of the microorganism's growth based on these algorithms.

We then faced the problem of how to recreate the correct conditions outside of the scientific laboratory, to encourage other people, designers or artists to work with unorthodox materials, all the while guaranteeing sterile conditions. Together with the Mediamatic foundation in Amsterdam, I designed a clean room, a very simple but fundamental project that I developed with other research institutes in Amsterdam to work with microorganisms in sterile conditions - an essential factor if we want positive results. It is a very simple setup in which a blower inhales the air from the outside. This is filtered by a pre-filter that removes the larger particles and is pushed up through two

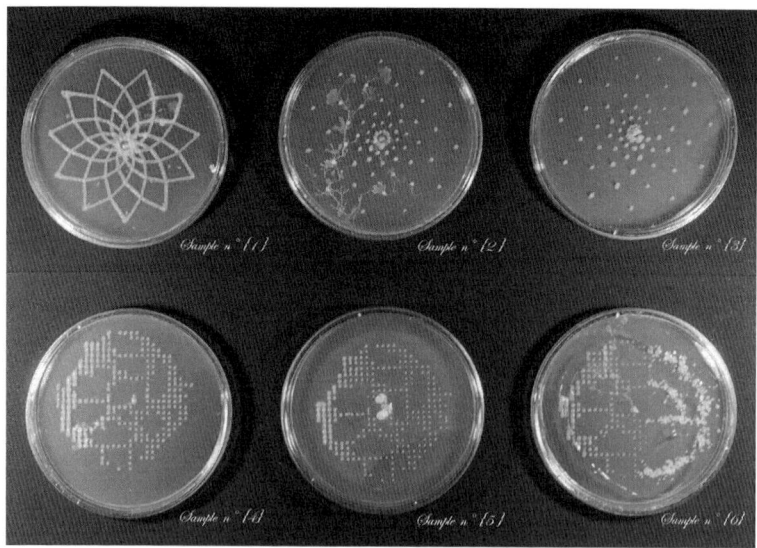

Fig.9 BioLogic timelapse overlap-frames. Courtesy Officina Corpuscoli.

HEPA filters (able to extract 99.98% of all microbes and the particles present in the air), leaving the air in the clean room perfectly clean and sterile and ready for monoculture work in conditions of absolute security [Fig. 10].

The interesting result of this project has been the creation of a community. If initially I was brave enough to embark on an idea that was defined "a little strange" by those who asked me what I was doing, and if I was an engineer, a scientist or an artist, following the project's execution, interest grew culminating in the constitution of a solid group of people and then of other groups, not only in the Netherlands but throughout the world. Groups of designers who believe in the open source method, in the distribution of knowledge throughout the entire biohacking circuit. They work to democratise the scientific means, or rather to obtain expensive tools within the scientific field with just a few hundred euro. It is a way of thinking that will influence the way in which certain technologies are used and taken forward. The public doesn't just watch, it also plays an active role and decides how to work together with the scientist. The decision becomes collective and more transparent.

I have also worked in other more conventional types of workshop, such as those in which we learn to cultivate fungi at home, so that people are attracted by the idea of growing their own food. A practice that also invites us to reflect on how to develop a procedure that is very similar for the establishment of alternative materials, which is the main aim. In one of these workshops we learnt how to pasteurise straw. It is a very simple process in which the spores of the fungus (we used grains that had already been colonised by the mycelium spores) are scattered and spread in a uniform way and distributed within the substratum of the straw. This is the nutrition for the fungus which recognises its highly elevated cellulosic content, and breaking the layer of lignin of the plant, it is able to colonise it by feeding off of the cellulose. Once prepared, this "mix" is enveloped,

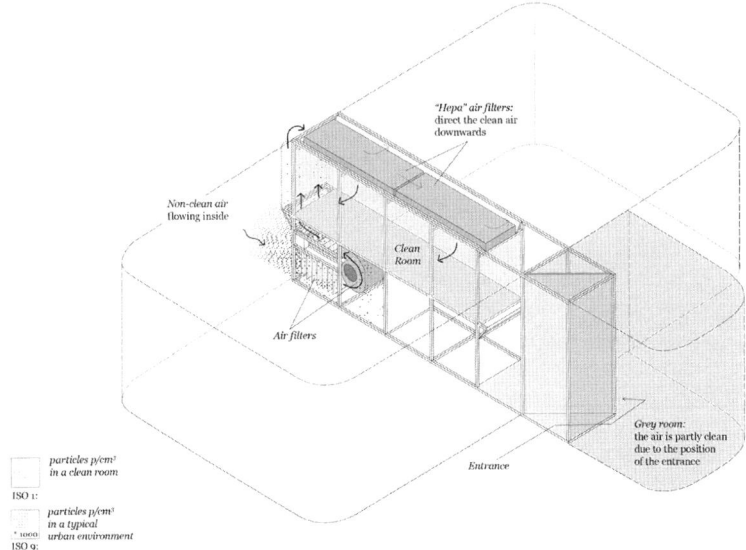

Fig. 10 Clean Room at Mediamatic. Courtesy Officina Corpuscoli.

over time the mycelium develops and the fungi grow as soon as the environmental conditions are favourable. This process is based on waste materials: agricultural waste such as in the case of straw, or waste left over from a manufacturing process such as sawdust; fibres, such as hemp, and any other type of plant wood-cellulosic material is a potential substratum for the fungus that will convert the cellulose into a sort of chitinous material.

In one of my first experiments, the substratum was made up of straw mixed with different types of hard woods [Fig. 11], in which we started building a primitive and organic form, a sort of "brick" produced thanks to the use of the fungus.

On one hand, the fungus consumes the cellulose therefore the sugars that it finds in the substratum and transforms a part of the material into chitinous matter; on the other side it acts as a glue, a natural resin that binds the fibres together, a sort of extremely resistant cement. This shows us how many advantages are available if we can replace chemical and toxic resins (such as epoxy resins) which are used in most building materials (e.g., MDF, and all composite materials that use resins) with alternative, natural and equally long-lasting materials, which have excellent mechanical and non-toxic properties.

It is also interesting to observe how the materials can change drastically depending on the substratum used, the growth conditions of the microorganisms and the type of microorganism used, even simply depending on the different strain within the same species. There are many parameters that may be used, and many results that may be reached. It is a very long process that I have been undertaking for some years now, which will have multiple results.

The course of my research is therefore based on these questions: what does it mean to produce? What does it mean to produce a material?

The crux of my reflection is the designer's ability to generate new view points and new production methods, with processes that are no longer based on conventional methods replaced by processes of growth and cultivation.

We could talk of a step back in an ancient practice such as agriculture which is updated again in the perspective of the present day and pushed forwards in a future vision in which materials and products will be the result of a growth process, therefore products cultivated as an alternative to the use of traditional plastics that use materials without negative environmental impact: preferable materials.

You will have noticed that I have never used the term sustainability. Insofar that the base of the projects illustrated here could be defined sustainable, I don't like using this term because of the over-use and abuse that has been made thereof also by "greenwashing" marketing, which has usurped its value. Sustainability does not yet exist, it is something to tend towards, to aim to build, but at the moment there is still no process created by man that has no impact on the environment.

For example, to make materials to replace plastic, such as materials starting from mycelium, or which in any case use elements in plastic, banally to maintain certain conditions of humidity. The realisation of something through a completely pure cycle and process: I do not believe this yet exists in human culture, even if we are trying to find it. So it is better to speak of materials with low environmental impact.

With my subsequent project I tried to outline an alternative domestic panorama, using a material cultivated within known forms that would produce very simple archetypal

Fig.11 Brick primitive. Courtesy Officina Corpuscoli.

forms that are able to bring better understanding of the possible applications for the material itself (sometimes even in a provocative way).

In the exhibition The Future of Plastic, which was held in the Plart Foundation in Naples in 2014 (despite the fact it was intentionally ambitious to speak of this as the future of plastic) what I have tried to demonstrate is that by eliminating any type of reference to fungi, no one would understand that fungi had been used: the usual reaction of people, in fact, is to talk of plastic [Fig. 12].

The materials used are intentionally primitive, specifically to highlight the organic language that characterises the process, but this does not mean that completely smooth and resistant materials cannot be created.

With the support of the Delft University in Eindhoven, I recently carried out some tests on some samples of materials whose mechanical characteristics are very similar to those of MDF, or which may be extremely soft materials such as polystyrene. [Fig. 13] I created a number of objects such as vases, lamps, plates, tiles - very simple items, so as to represent the broadest range of fields of application, from product design to architecture or even fashion, with the development of materials that may replace fabrics. All these materials are currently in the test phase. We are collecting data to fully understand the characteristics of and protocol each individual growth process of the specific material in order to identify a standard process that will safely lead to the development of new paradigms on a large scale.

This is, in fact, my current goal: to reach a certain scale, not for personal ambition, but because it is vital that these projects have an impact on the social level and not merely remain valid on the speculative level. They are tangible projects, but they remain in the circuit of galleries and museums, in the cultural niche, and they do not create social impact. To have real impact, this alternative paradigm must affect people, and this can only happen if we can start collaborating with industry.

One solution could be to do my best as a designer: not to wait for the industry to catch up with me but to create the industry itself. This idea currently finds me working on creating a group based on existing industrial realities to produce this type of material on a wide scale alongside the food industry of fungi, which is a very high level manufacturing reality [Fig.14].

What I have spoken of until now can be defined as composite materials, in which the fungus acts as a binder transforming a part of the substratum of the agricultural waste into chitinous matter. These contain a dead plant element.

Moving on, now I am interested in the production of completely pure materials which are exclusively made up of mycelium and characterised by the same chitinous

Fig.12 The Growing Lab - Mycelia. Vessels overview from "Future of Plastic" exhibition. Courtesy Officina Corpuscoli.

substance as fruit itself. This is therefore the direction in which I am taking my research within Utrecht University alongside my scientific partners. Pure mycelium for example could compete with plastic materials [Fig. 15]: it is in fact a good alternative to polystyrene, a potential non-woven textile (a fabric whose fibre is randomly rather than perfectly distributed), or it could have the consistency of wood [Fig. 16] or extremely flexible materials [Fig. 17].

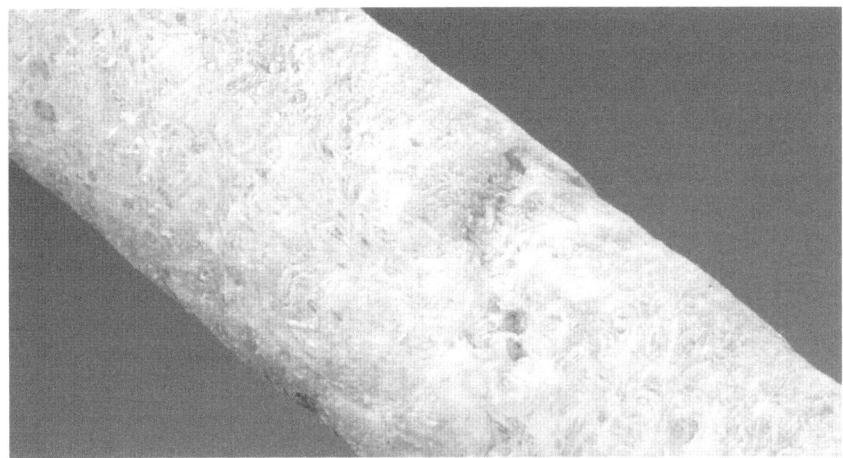

Fig.13 The Growing Lab - Mycelia, pure mycelium sample soft absorbant.
Courtesy Officina Corpuscoli.

But the laboratory is not an industry. It is a place in which you try to understand which strategies will best lead you to define a precise protocol in order to be able to replicate the experiment in exactly the same way and calculate the costs and advantages, as it is naturally difficult to compete with widely-distributed materials and established processes, such as those behind the production of polystyrene.

Fig.14 The Growing Lab - Mycelia, mycelium vessels overview.
Courtesy Officina Corpuscoli.

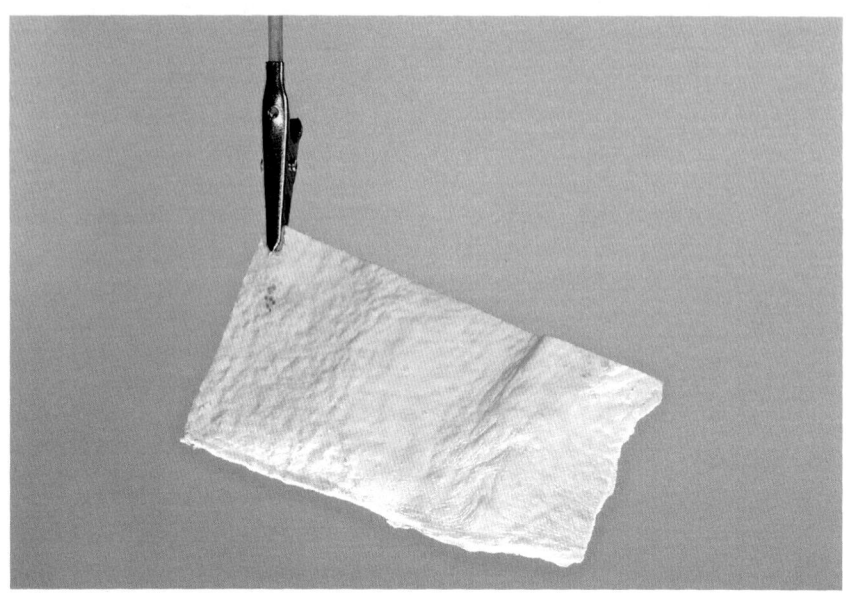

Fig.15 The Growing Lab - Mycelia, pure mycelium sample flex textile.
Courtesy Officina Corpuscoli.

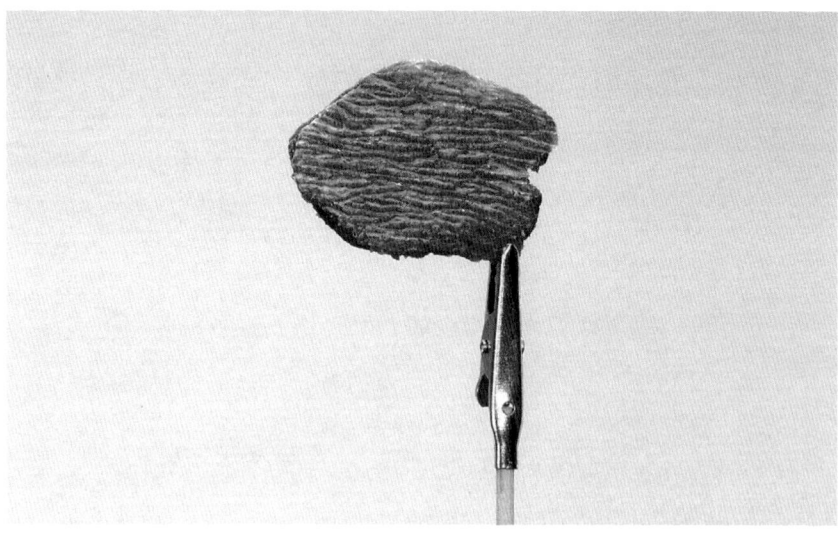

Fig.16 The Growing Lab - Mycelia, pure mycelium sample rigid plastic.
Courtesy Officina Corpuscoli.

At the moment, we are trying to improve the growth conditions of fungus, experimenting with different species and developing the genetic level: a debatable approach even though in this case the plant matter is "killed" once the growth process is terminated, for which the final material is completely inert.
Once more however we are faced with ethical dilemmas which require debate.
In conclusion, my research mainly focusses on the importance given today to the analysis of processes and cycles of transformation, in that everything that exists is (or is presumed to be) "matter" that can be transformed, change its own characteristics and become something else: another form, another type of use - and we may just find that fungi, in this sense, are of considerable use.

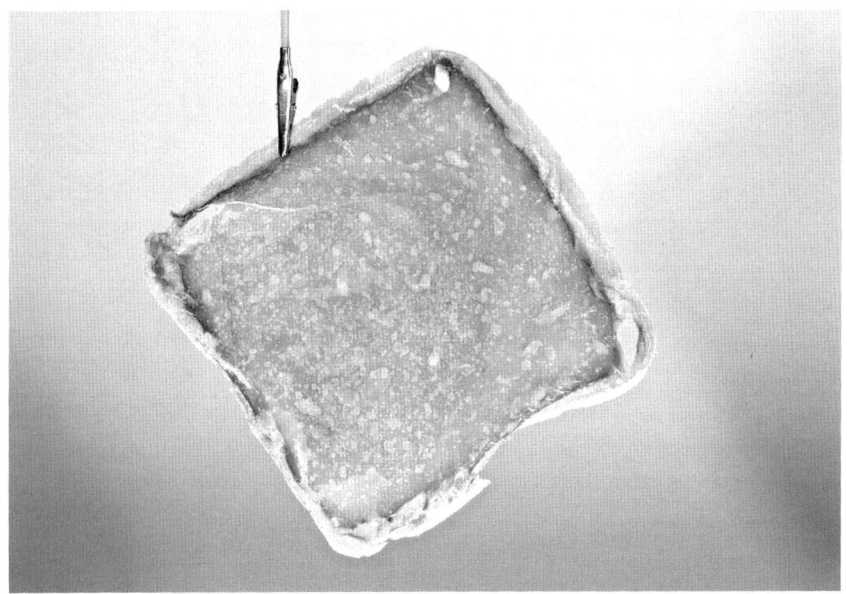

Fig.17 The Growing Lab - Mycelia, pure mycelium sample flex plastic.
Courtesy Officina Corpuscoli.

References

Montalti, M. (2010). *Continuous Bodies: cycles of decomposition triggering a symbiotic partnership between humans and fungi*. Final project for the IM Master Course - Conceptual Design in Context. Eindhoven: Design Academy Eindhoven

THE ADVANCE FROM THE EXTERIOR TOWARDS THE INTERIOR

FORTUNATO D'AMICO

The start of this millennium has undoubtedly been characterised by the technological revolution, drastically changing the orthodox perspectives through which for centuries we have observed and studied the reality around us.
The conception of relativity, meant as the base condition of the principles that support modern sciences, has moved our model of thought, aiming it towards infinite horizons and escape routes compared to the axonometric-orthogonal ordinary of modern day, fluidifying, in fact, the spread of information, nomadism, and economic and productive globalisation.
The "internal" vision has broadened our ability to perceive and encompass the "external". Among the positive aspects, inspired by the idea that each subject is understandable and gains value only if related to other exterior factors, is also that which has allowed some contemporary theorists and scholars, the task of promoting a return to an interdisciplinary character in all sectors of science and culture.
In the education field, the bringing together of knowledge that were separated before may be the perfect way to implement the levels of participatory democracy and the processes to share the transformation of the sensitive system that we set about modifying and governing in this five-year period of the century.
Explorations have affected the notions and current terms of use, such as those of "internal" and "external", often inverting the established functions. In this way, it is today possible to dislocate our activities and knowledge within the networks system and therein manage our working and social relationships on a global level.
If we think about the planet being covered by invisible wires, cabled to quickly transport information all over the world, this leads us to intend a dimension of the "internal" which certainly disorientates and makes obsolete the previous way of understanding it, taking a significant step forward in the metabolisation of the principles that have, until now, ruled the disciplined of architecture and design, and more in general the planning of the natural landscape and industrialised production. It is, therefore, opportune, every time we speak of the internal-external dyad, to establish the basic conditions to attribute the appropriate characteristics to these categorisations.
How many interiors in an interior?
How many exteriors are there then in an exterior?

A close examination of the details in what we observe will let us understand the "overall" like a multitude of functional elements that are integrated amongst themselves and operate like fractals.

The danger faced by contemporary design disciplines and schools is that they underestimate the presence of this new awareness that was born by the computer revolution, continuing to commit the catastrophic error of culturally forming intellects that are specialised in just a few centimetres of knowledge, avoiding contaminating them with the knowledge of any similar disciplines.

Michelangelo Pistoletto, intervening in this course, promoted a transversal and relational thought in which two subjects, conceptually and physically in contrast, generate a third element born from their interaction. Dynamic, Trinamic, Third Paradise – these are the pragmatic axes of the theory of transformation and the completion of the work of humanisation which has already been kick-started by mankind and now in the progress of being perfected passing through the turbulent start to this millennium. Michelangelo Pistoletto believes that we are heading for an era in which we will have to find a new equilibrium, returning to a dialogue with nature and ourselves.

The idea of a strict interdependence between things and people allows us to access a deeper understanding of the phenomena that surround us and to organise crucial policies compared to the problems presented, in order to safeguard the presence of human beings and other biodiversity on the planet.

In this sense, the understanding of our life within the terrestrial biosphere - which is in turn surrounded by astronomical phenomena that determine the conditions for the cyclical renewal of our existence on this planet - cannot but consider the dimension of our presence "within" the system of cosmological and micro-cosmological relationships.

The inattention paid by contemporary culture to celestial phenomena and their concrete action on the evolution of everyday life, was caused by an excess of extreme artificiality of today's habitats.

The conditions of modernity have narcotised the vision of reality, distancing nature and agricultural production from residential areas, leaving space to the indiscriminate overbuilding of the territory.

Originally, cities were founded "within" natural, resource-rich places.

Over the millennia, populations tried to keep the equilibrium between nature and the artificial stable, transmitting afterwards a philosophy and science characterised by dialogue, morally condemning the abuse of men over men and artefacts compared to the environment.

The contemporary metropolis, on the other hand, has chosen to infringe these rules, accelerating construction growth and thereby annulling any chance of productive and poetic debate between the inhabitants and nature from the territories.

In this third module of the Interior Design course students have investigated, studied and designed visions of a future that considers the delicate mechanisms that unite the exterior with the interior.

Within this logic, we have realised the importance of the designer's role "within" the complex productive system to which the redevelopment of the planet is entrusted today.

The responsibility of the designer and his products towards society and the environment for which he works, must remain clear, especially today that we are witnessing helpless the continual growth of waste in tips in urban suburbs or placed in strategic places throughout the planet, where every second tonnes and tonnes of rubbish are accumulated; rubbish which, before becoming such, was consumed as design objects and has now become pollution, poisons that are impossible to dispose of, becoming the symbol of unrestrained consumerism that has subjected the discipline to the commercial needs of manufacturing companies without a thought for the final client and his habitat.

RESEARCHING THE FUTURE - PROFESSIONAL EXPERIENCE

STEFANO MARZANO

1. INTRODUCTION
I started working as a designer in the 1970s. My professional experience with the Philips company lasted twenty years, and covers over 60,000 projects. Other experiences followed, such as courses in the Domus Academy, in the Politecnico di Milano, and more recently, the THNK School of Creative Leadership founded in Amsterdam in 2011, various activities for the European Community and finally, over the past few years, for the Electrolux Group.
At its height of development, Philips Design employed 600 people, offered services and acted as an incubator for other companies with non-conflicting business interests. Philips Design represented a business unit, a company within a company that worked for Philips but also for other businesses, not only to generate profit to be reinvested in research activities for the business unit itself, but also to explore other corporate cultures, other technologies and ways of viewing the market and the consumer, so as to enrich Philips' culture of project and innovative vision.
Here, I would like to illustrate some projects that have played a significant role for Philips Design and for design research in general. The majority of the research projects that I directed have already been published. The decision to publish unfinished projects is a huge step forward, as large international companies tended to keep the results of their research secret – even within the company itself - in order to avoid the leaking of information to competitors. The reason we have decided to publish the research comes from our conviction that this passage was functional for the research itself. Ours was a speculation on possible future hypotheses of which we had no absolute certainty, in that we were not actually certain of society's "desirability" of what we proposed. The publication of the projects helped validate the hypotheses put forward. We started to involve a wider number of people: the *Vision of Future* project in 1996, for example, used the Delphi method[1] and the participation of various professionals and

---

[1] The Delphi method is a structured communication technique or method, originally developed as a systematic, interactive forecasting method which relies on a panel of experts. The experts answer questionnaires in two or more rounds. After each round, a facilitator or change agent provides an anonymous summary of the experts' forecasts from the previous round as well as the reasons they provided for their judgments.

researchers in a variety of disciplines to which we subjected our concepts to obtain their opinion regarding the probability that the hypotheses could become reality and on the desirability of these "hypotheses of a possible future". The *Television at the Crossroads* project from 1994, developed with Andrea Branzi, Alessandro Mendini and young designers from Philips Design, was conceived for development in three "art studios", or rather three groups of creative people working separately each under the direction of Branzi, Mendini and myself. The project explored the convergence between the television and the computer in the emerging digital world, configuring visions that would meet in applications which today seem almost banal, but which at the time were extremely innovative.

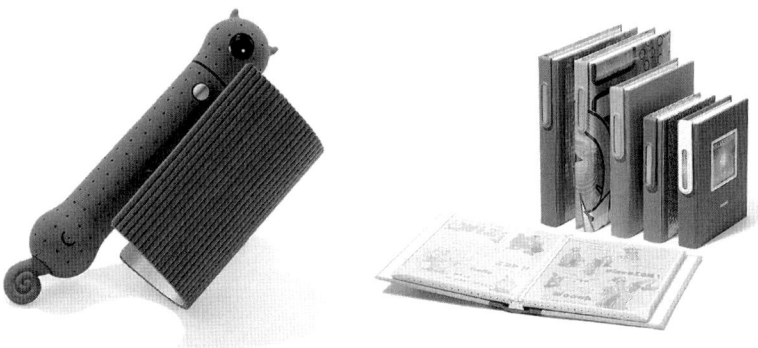

Fig.1 Philips Design, Vision of the future, 1996.

2. HIGH DESIGN

When, in 1991, I was called to manage the Design group, Philips was celebrating its 100th anniversary. In 100 years of history, Philips had transformed greatly: from focussing on technology, to international growth and the development of global markets and thereon the development of sales organisations in sixty countries throughout the world.

We should bear in mind that in the 1970s, companies had to deal with a saturation of the markets and with a reality that no longer favoured companies that generated new technology in the production-sales process. We were facing a new problem: we had to understand the needs and qualities that the world needed. It was about adopting new innovation and marketing strategies and generating a strategic overhauling.

At the end of the 1980s, Philips found itself in a difficult and complex situation, nearing bankruptcy, due to the change in the global situation and the pressing competition of large Japanese brands.

When I joined Philips, I was lucky enough to have a very direct interview and build a good relationship with the company's chairman at the time. I made him a rather unconventional proposal.

I had been very impressed by the work of Roberto Assagioli, an Italian psychologist who had developed psychosynthesis and had theorised the multi-personality, or rather

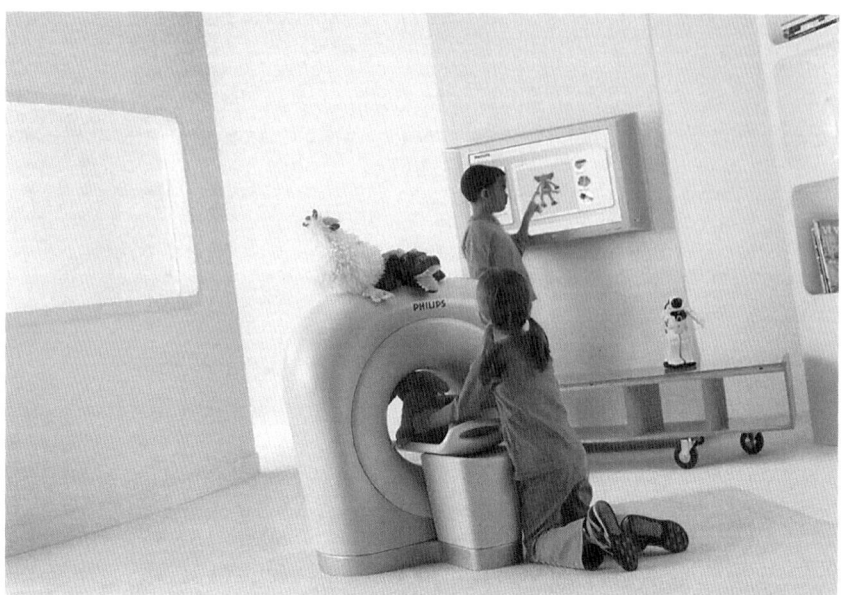

Fig.2 Philips Design, Experience Design, mini-scanners.

the theory that each person has a number of personalities: the personality of the hero, that of the frightened child, and so on; different personalities that our Ego keeps under control, trying to discipline the parts of the "bad" personality and focus on the "good" personality, or whatever we would like to be.
I proposed to adopt this theory after observing Philips' reality in the 1970s, in which the company's culture was going through a period of transformation following the exit of the Philips family and the start of a managerial culture which focussed solely on short term commercial values. The challenge at this point would have been to find the company's Ego and bring out its better personalities, those that were devoted to the wellbeing of the company and innovation.
Following my proposal, the history of the company was reviewed, including its inventions and innovations, in order to understand what the company's foundations were. This operation meant that the essence of Philips could be brought back to the surface and shared once more – the value of research and developing technology in order to generate products and solutions to improve people's lives: "Technology to improve the quality of people's lives". It may seem banal, but it encompasses the concept of research to understand what is really necessary and how we create it. At this point, it was necessary to understand how the markets had changed, how the company had changed and how the target consumer had changed and also rethink the relation interfaces with the market and everything regarding distribution. Today, for example, we talk of *experience* and *touch points* that the consumers may have with a brand or a company.

During those years in the company, I developed *High Design*, better known today as *Design Thinking*, a strategy of changes and transformation of the creation process within the company.
The first point of the strategy was to understand the consumers' lives, then to analyse the socio-cultural, economic and geographic contexts. The second point was to work with the socio-cultural research tools rather than those of marketing – superfluous and superficial – to better understand the lives of people, their values and aspirations. The third point was to integrate various skills within the creative process, avoiding too many sectorial design activities and creating large interdisciplinary project teams.
We first introduced this strategy in the Design structure, and then shared it with all the business functions, spurring communication between the players of the corporate system: not in competition, but with an eye more towards collaboration with a modest attribute to show that we designers are no different from others, we also have our limits and we need to cooperate.
This is the summary of *A new strategy for Innovation* from 1992:
1. Centred and focused on people
2. Research based
3. Integrated multi-skill
4. Integrated in business process

The name *High Design* intended to cause a stir within the company in order to "elevate" the concept of design to a higher level than that previously known: to a level of high social responsibility, humanist culture and high technological knowledge.
In 1991 and 1992, when there was still no talk of sustainability[2], I supported the concept of "etica nova", or rather a humanist-renaissance type vision with man in the centre of the world, but also a concentricity between the human sphere, nature, artificial landscape and everything else necessary for quality of life.
Leonardo da Vinci, the Renaissance intellectual who encompassed astronomical, mechanical and artistic knowledge, inspired the model of interdisciplinarity. I thought a new Leonardo was impossible, but that we could create an organisation that could be a "Leonardo". And that a team built on common values, made up of various disciplines that collaborate, at the same time share problems, needs, challenges, could come up with a genial thought: the ability to think beyond convention.
I presented my idea of the new organisation with a design, which shocked the corporate management. It was an organigram that was rather complicated as it represented a non-static but organic organisation. The design of the organigram was similar to an animal whose head is the part that unites, manages and coordinates.
According to my idea, the Design Competence section dealt with skills development. It was made up of people who worked on skills development programmes, a sort of university within the organisation. The Design Business section, on the other hand, had the role of interfacing with all the various product divisions, of which there were

[2.] The term "sustainability" took on importance after the United Nations Conference on the Environment and Development – UNCED – in 1992 in Rio De Janeiro.

twelve at the beginning (today there are only two and a half, the same company has gone from 440,000 employees to around 80,000, following the split of the Lighting sector). Finally, the central part of the organigram, the section named *Design Fluidity & Knowledge Transfer*, accounted for the majority of people.

The organisation's main principle was fluidity, being able to easily move people from one activity to another, thereby creating interdisciplinary groups that would mutate or redefine themselves depending on the projects. Every project was evaluated to understand which abilities were necessary to develop its contents, after which the project teams were set up.

The theme of the Design Research activities was to "create" visions of the future to explore what could represent a "new value" for a desirable future life. The second step was to develop the solution hypotheses, which required validation, and above all to ask which skills, within the company, would be necessary to give form to these solutions. This was a very important aspect. To create a "new value", it was necessary to also have a critical position regarding what knowhow one had within the company. Therefore, asking oneself if further skills, new collaborations or acquisitions were needed. All of this is normal today for international companies which practice "open innovation" but, at the time, there was a certain closure against the world outside of the company. It was therefore essential to accept that not all skills could be integrated within the company.

Other important points included the quality and intensity of the relationship with the consumers and clients in the research activities. Personal relationships are today recognised as being important in determining the type of experience we want the final consumer to have. It was, therefore, necessary to understand that people should be analysed in their socio-cultural and socio-economic ambit according to three different parameters:

1. an ethnographic parameter to create relevant insights for the short term;
2. forecasting activities that through the understanding of cultural changes attempt to identify the aesthetic attributes that can give meaning to the new value;
3. foresight activities, which deal with understanding the emerging social dynamics and to identify which ones will be dynamic in the long term, and able to influence the future in a determining way.

On these three levels of research, and with the involvement of the sociologist Francesco Morace, who I met at the Domus Academy, we set up the *People & Trends* programme divided into the various geographical regions of the international market.

The model that I had proposed to top management was, in brief: to create accelerators for innovation.

3. RESEARCH AND VISIONS

One of the most interesting research projects undertaken in Philips Design is, in my opinion, *The New Everyday* (2003) on the theme of intelligent environments, in which technological research and design research have been aligned and integrated. This research was published in a book that was the first to show concrete projects for new applications of information technology and smart technologies of great potential. It gave the sense of what we could create, giving sense to what had already been made, but without us fully realising that we were talking of "intelligent environments".

In the *Vision of the future* research (1996) new concepts and applications were created and illustrated in a series of very realistic futuristic films. These films, seen again today, seem to be stories taken from the reality of today's daily life. Nothing new. However they tell of a vision that twenty-five years ago had a very important anticipatory meaning.
In occasion of the publication of *Vision of the future*, an exhibition was also set up inside Philips: a provocation to Philips itself that aimed to spur another debate on its products portfolio and market but above all aimed to inspire reflection on the "new value", its meaning and form.
The concepts (objects, interfaces, interactions, narrations and film) were divided up into four environments: *personal, home, public and mobility*. The choice was intended as a way of segmenting life on the axis of the space and simplicity of the matter. Materiality relates well with us as we are physical and material, and therefore it is more easily understandable.
Formally, the concepts did not adhere to a technological aesthetic, such as that of the most popular products of consumer electronics: the television or "black box" and then the "grey boxes" - the first computers. The concepts instead had warm, homely qualities and interfaces that were equally friendly and homely [Fig. 1].
Research has explored new applications through objects that put themselves forward as "transfers of emotion", using wearable technologies as well. An experimental activity that once more aimed to provoke and redefine the company's strategy and its business portfolio rather than propose real new products.
Another subsequent exploration entitled "New objects, new media and old walls" clearly communicated the concept that the transformation of spaces would have been stronger within homes that in cities. In actual fact, the huge change was that within the places where we spend our everyday lives, the old walls are those that we know well, yet are those that would quickly see the addition of new objects and new media. Objects that expressed in any case a domestic quality, such as for example rugs made in connective tissue to recharge mobile phones, or the miniscule projectors that fling interactive images onto walls and ceilings. With *Vision of the Future* we started to become aware of what information technology would mean in our everyday lives. Today it is difficult to understand how people lived in 1995 without today's technological benefits. Mobile phones looked like a large box with a handle. Speaking of the "volatility" of technology was, therefore, a novelty. Today, on the other hand, we live alongside technical products with great simplicity and naturalness, as if it has always been this way.
The message was then that the present would be a home full of "black boxes" and "grey boxes", that the future would have been much more similar to the past than the present. Because new technologies, which can be miniaturised and intangible, have allowed the integration of new communication functionalities in all objects of cultural and anthropological importance within the spaces of our everyday lives, respecting the diversity of cultures (what is commonly known today as "the internet of things"). If mats were more important than chairs – like in the Arab world, for example – technological functions would have been integrated within them, rather than in a chair.
The project included objects such as the bookcase with scanner and integrated electronic books, interactive objects linked to the culture of food, objects for commu-

nication in the home – between the dining table and the kitchen – all wireless objects such as the tablecloth made of conductive fabrics connected to supply energy to other interactive objects found on the table: LED candles, the audio player system, the telephone, etc. - everything without electric wires that still today burden our homes.

The *Ceramic Audio* project, which only reached the prototype stage and was created with Alessi, was an attempt to translate this research into a collection of products that could be launched onto the market: it is important to launch a five- to ten-year vision to understand in which direction to move even in the short term. The objective was to create new audio products integrated in ceramic objects. We had to halt the project due to the complexity of distribution, logistics and volumes for a collection of this kind: industrial electronics within artisan ceramics. A part of the organisation was still incapable of innovating the distribution strategy and logistics.

The products of the *Plugged Furniture* programme designed with the Leolux company were introduced in the Northern European markets. They proposed the integration of the "black object" in striking pieces of furniture following the above logic.

A similar project, on the integration of different technologies inside a seating system, was undertaken with the Felice Rossi company. Lighting, audio and multimedia interactivity were integrated in modules inter-connected with electromagnetic contacts. The modules offered a network of connections for high performance objects and seating variations. *Paesaggi Fluidi* is another similar project undertaken with Cappellini, for which a number of furniture products were created with integrated sound and music functions.

4. AMPLIFYING THE ABILITIES OF THE HUMAN BODY

Inspiration struck after reading the book *Le Geste et la Parole* by French anthropologist André Leroi-Gourhan. What impressed me about the book was the narration of the human journey and the idea of man committed to the exteriorisation of his functionality in tools, machines and objects of common use with the aim of increasing their effectiveness and amplifying their capabilities.

I tried to translate this thought into a diagram in which exteriorisation first passed by simple and small objects which could be held in a single hand to increase their capabilities, right up to objects that contained the body itself (carriages, cars, airplanes) - therefore objects able to increase the ability to move. All while I asked myself what would have happened in this process with the impact of "volatile" technology and extreme miniaturisation. I recognised a possible reintegration of some human functions: an exteriorisation of the functions that have to consider our materiality and the re-interiorisation of all other immaterial functions. In particular, I was interested in thinking about memory and communication. From here, the definition of "interactive tattoos", "interactive piercings", chips to insert in teeth and under the skin. All of this gave me an idea of human ambition. That of aspiring to having the divine qualities of "omnipresence, omniscience and omnipotence" attributed by man to the Divine. Or rather the desire to be divine! With all the relevant benefits and risks. The means or tools in use have neither positive nor negative qualities: it is the hand that uses it, the brain that commands it that turns it into a tool of life or death.

Another provocation of the *Vision of the Future* was dedicated to wearable technologies. Initially they hadn't been studied in depth until a project team from MIT presented an

interpretation of wearable technologies which for me represented everything that was not to be generated. Thus, we were encouraged to rethink them.
The reaction to MIT's interpretation was the *New Nomade* project, the world of wearable technologies as we had imagined them in *Vision of the Future*. Our vision was that technology should not be seen as integrated within clothes, but that they would have to be secondary compared to garments. They must be present, but invisible, so as to provide their services in a magical way.
The first step in translating this research into reality and development was a project undertaken in collaboration with Levi's. Speaking of wearable technologies was fascinating, but it also gave us some considerable challenges: integrating electronics in machine-washable clothes was not particularly simple. Firstly, we had to deconstruct the functions (telephone, player, mp3, etc.), integrate them in the clothes in a comfortable, simple and elegant way, and create internal connectors and circuits that linked them by function.
I always thought that the most sensible application of wearable technologies was during sports activities, when you need to have your hands free. This led to our cooperation with Nike. Initially, we developed wearable audio products for sports such as jogging and climbing, extending our goals to the creation of integrated applications for emergencies such as illness during sport , and for calling emergency services in the event of attack and danger which could be applied while jogging in city parks at nightfall.

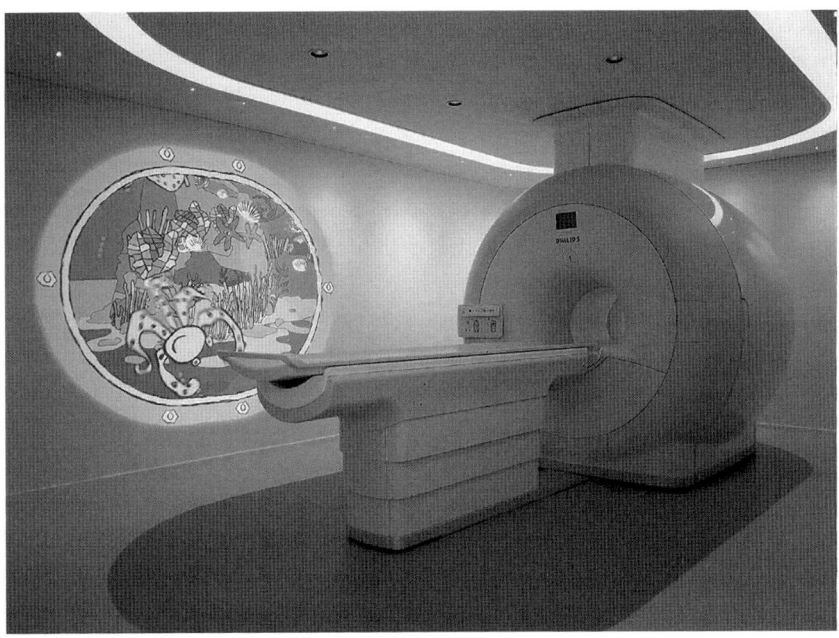

Fig.3 Philips Design, Experience Design, ambiente experience.

From here we started imagining the connected athlete, one who wears clothes through which, for example, he can send biometric information to allow the monitoring of his vital functions. To then extend the aim to the use of data collected regarding athletes and video recordings made with cameras integrated in sports uniforms, which paved the way for new and completely personalised domestic televisual entertainment. Television production at the service of the viewer - giving him the chance to choose the best views during a race or a match, or even interact with it directly in real time.

5. EXPERIENCE DESIGN

In the medical arena, our research generated the transformation from *System Design* to *Experience Design*. I am talking in particular of a project that represented an important moment in this journey: *Paediatric challenge in CT environment*. The project was started in a paediatric hospital for chronic diseases in which the children were subjected to treatments in environments dominated by machinery and technology. Environments which caused anxiety in the children and their families.

The desire to reduce this anxiety and to create environments where the youngest patients could be more at ease led us to create a new design process, which we then called *Experience Design*.

This design process involved medical staff representatives who weren't involved in treating the patients. Subsequently, we then involved the patients and their families in the design as well.

We created a series of images that focussed on each moment of the medical examination highlighting the emotional aspect of the experience.

The first project team was made up of an interdisciplinary group in which architects, designers and doctors no longer had to work on defining an individual product, but on redefining an entire medical ward.

The elements involved in the new experience within the hospital included the use of projections, the invention of hypnotic stories and the creation of educational messages. One of the biggest problems was the reaction of the children subject to the medical exams. In many cases the children suffered serious anxiety crises, meaning it was necessary to sedate them in order to undertake medical exams such as CT scans, which in normal conditions would take twenty minutes but when anaesthetic is required can take up to two hours in all.

This shows how important it is to distract and calm the children, so as to reduce examination times and treat more patients, but also to obtain better results from the psychological point of view compared to the experience of the children and their families at the end of the medical examinations.

This was the starting point from which we completely re-designed the process of medical examinations. Firstly, one fundamental aspect was the information of the patient, so we designed some mini-scanners [Fig. 2] which simulated through play the experience that the child would have during the examination. Then, in collaboration with a team of paediatric psychologists, we decided to create some hypnotic stories for the children to concentrate on thereby drawing their attention away from the examination.

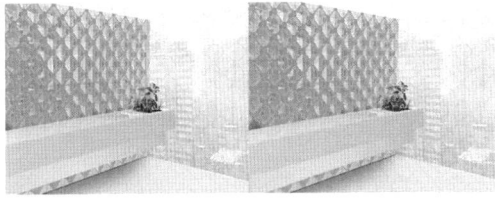

Fig.4 Philips Design, Off the grid: Sustainable Habitat 2020.

Another important aspect of a positive examination experience was eye contact between patients and medical staff during exams. As he approaches the machine, a child must feel that he is in a friendly and familiar environment.
The diagnostic rooms could be transformed with different atmospheres created through the use of ambience lighting and hypnotic stories to distract the patient and help him to relax [Fig. 3].
The final results were notable: 25% of children did not require sedation. This is a very important result, especially because in the United States, where hospital specialisation has become increasingly important in the branding strategy, hospitals are now judged by their competitive improvement in the wellbeing experienced by their patients.
Finally, I would like to introduce some research on global dynamics, in particular on the question of over-population and the relative changes in the concept of "living". Studies on this theme mainly regard Asia, a geographical area in which cities change size at an unbelievable speed and over-population reaches extremely high levels, unlike European cities, where changes occur mainly within living spaces.

In China, where new cities are founded and expand at a fast rate, we need a view on how this growth may occur, not only from the town planning perspective, but also from that of design which views buildings as a high-performance machine-product of new living qualities, an organism of habitation.

In this case, the provocation was to copy nature, and think of "living" as a tree, for example, which inhales dirty air and exchanges it for purified air, which collects rain water and purifies it, which filters and transforms waste water and reuses it, which captures natural sunlight and distributes it in the environment via filtered facades. The *Off the grid: Sustainable Habitat 2020* [Fig. 4] exploration thus dealt with the future of megalopolises in China. A project that is still of great interest today and which, in my opinion, will mark the future of cities and the future of humankind.

References

Aarts, E., Marzano, S. (2003). *The New Everyday: Views on Ambient Intelligence*. 010 Publishers.

Branzi A., Marzano, S., Mendini, A. (1994). *T elevision at the crossroads*. V+K Publishing.

Philips Corporate Design. (1996). *Vision Of The Future*. Eindhoven, Netherlands: Philips Corporate Design.

CITY-WORLD AND WORLD-CITY

MARC AUGÉ

It is difficult nowadays to distinguish the "exterior" from the "interior", the "beyond" from the "here".
The architect and philosopher Paul Virilio theorised this aspect of contemporaneity in an attempt to define the character of the system of globalisation, starting by analysing the strategy of the Pentagon. In his book *La bombe informatique* Virili (1998) o sustains that globalisation has flipped the relationship between inside and outside. The interior becomes "global": the system, or more precisely the economic and electronic networks that make up its concrete foundation. In this vision, "local" represents the exterior insofar as it claims to escape from the system's logic. It is a provisional exteriority of the perspective of the system, both because the local is shortly condemned to take on a global appearance uniting with the logic of the market and the communication networks, and also because its claim for exteriority gradually - and in some cases, brutally - removes it from the pressures of the system, as so beautifully summarised by the phrase "right to interfere".
Paul Virilio was the first to coin the keywords of Western and contemporary global society: ubiquitous and instantaneity. These words describe an ideal and a reality that starts to look like that ideal. We are witnessing a multiple decentralisation that overthrows the relationship between interior and exterior: the decentralisation of the city, homes and individuals themselves. The urban space expands, and the frontier between the here and the there disappears. So-called "urban suburbs" extend farther than historical city centres and the latter are increasingly frequented by tourists from other countries. We are therefore witnessing a two-fold globalisation of city centres that become recognised and recorded on the world maps, diluting the city in an urban space without a clear frontier: a notable example of this is Paris.
In parallel, the effective context of large metropolises - multicultural and multiracial following migration and with all the inequalities due to the expansion of capitalism - becomes the planet itself.
Through my writing, I have tried to explain this evolution comparing the city-world - with its contrasts, diversities and inequalities - to the world-city which corresponds with the effects of uniformity created by the market and, even more, by the networks of global communication and the proliferation of images and messages. The world-city becomes the context of the city-world.
Today the true confrontation is between the world-city - which appears transparent and clear as it is so rapidly travelled through by world leaders, tourists, architects and

some professors; whose image can be found on all TV screens and travel agency display windows and which appears to currently be undergoing aesthetic and economic standardisation – and the city-world, the megalopolises where we can find all social and cultural differences, where poverty exists alongside opulence, and where just a few kilometres separate the laboratories of scientific research from the districts of the illiterate.

The world-city is the context of the city-world. The second decentralisation is that of the home, the most spectacular expression of this enterprise of communication networks. Jean-Pierre Vernant, the ancient Greek anthropologist, analysed the traditional Greek home (Vernant, 1990): the focal point was the hearth, dedicated to femininity and the goddess Era, while the threshold was the realm of the god Hermes, open to the outside and dedicated to trade.

Today, the decentralisation of the home has had a striking effect: thanks to the internet and television even in our private homes we are bombarded by images from all over the planet. Hermes is no longer on the threshold but in the centre of the home: Hermes has substituted Era.

The individual himself is armed with electronic prostheses that extend his body allowing him to receive and transmit data and enter an infinite communications network.

The city-world is a place, and this place is the measure through which we can interpret the social structure. The definition that I have tried to give is that of a place through which we can read the social structure and the history of a community.

The city-world, the megalopolis where we can find poor districts and rich ones, is the definition of the place itself. The world-city, on the other hand, produces an image of fluidity as we enjoy an air-show in the sky or watch television screens or surf the internet; in clear contrast with the rigid reality of local and real social life.

The change of scale and new contextualisation are the essential factor of our times. They are the deep cause of diversity. As urbanisation extends between the city and the countryside, any frontier between countryside and city is torn down: the urban culture extends. New contextualisation beomes essential.

On the subject of landscape, if it were possible to say that landscape is all culture, the natural landscape would no longer exist: in fact, landscapes always reveal the concrete or imaginary intervention of man. Moreover, today we are facing the problem that the planet itself has become a landscape. It is interesting to note how the consumerist society has made it possible to travel beyond the earthly orbit so that some – extremely wealthy – tourists may see the planet from above. I believe that it is important to highlight this evolution, as now the change of spatial scale also means a change in time scale.

Having started exploring our galaxy, we now know that our galaxy is home to billions of solar systems just like our own and that within the entire known universe, there are billions of galaxies. We cannot even begin to imagine these dimensions.

Using the planet as a landscape, we can project ourselves onto a still-uninhabitable exterior. But given that *homo sapiens* is merely 300,000 years old, its young age leads us to believe that there is still much to imagine about its future. It could be said that we are at the end of the prehistory of mankind: the history of planetary society starts now.

In conclusion of these reflections, here is a brief summary of everything discussed so far.

The first point that I dealt with is the presumed opposition between place and non-place. The place does not oppose non-place like good opposes bad. The absolute place, where we can interpret the social structure, is a place where everyone is under a kind of house arrest regarding age, sex, marriage status, residence. On the other hand, the non-place is absolute, a space without otherness, a space of infinite solitude. The absolute place is totalitarian, and the absolute of the non-place is death.

The second point I dealt with is the theme of the planet as a landscape. It is a radical and definitive change manifested by the end of the classic opposition between countryside and city.

The third point concerns the question: how can we create a place today?

We create places continuously: for example when we meet other people, our friends. The problem is that collectively we cannot create new places but heterotopias - using Michel Foucault's definition - or rather concrete, realised utopias (Foucault, 1966). The context of these heterotopias is always the non-place that can be defined as the context of any place. Today non-places, communication networks make up a system.

The last point takes into consideration the opposition between one and the other, between us and others. There is no identity without relationship, and a relationship requires time and space: relationships are not possible in the world of ubiquity and immediacy. These are threats against identity itself.

Today the symbolic is a form of resistance against the totalitarianism of the code. We may also think that the final confrontation is that between time and the history of men. In conclusion, I would like to talk of a utopic ideal: education for all.

We are living in a world in which there are huge differences in wealth and culture. The problem is that this difference increases every day, even in emerging countries. It is vital for us to redefine the necessity of a resistance. And this occurs through education for all. But, we are also well aware that this is a utopia that we will only ever be able to partially fulfill. Education could be defined as a heterotopia of any society. Education for all is a utopia, but it is a necessary one, without which nothing is possible.

Using this summary as a starting point, I would like to say something about the future: starting from the concept that place and non-place are perfectly solid and that the entire context is planetary, everything we can try to create is merely a temporary utopia.

References

Foucault, M. (2004) [1966]. *Utopies et heterotopias* [*Utopias and Heterotopias*]. CD :INA, Mémoire Vive.

Vernant, J. (1990). *Myth and Society in Ancient Greece*. New York : Zone Books.

Virilio, P. (1998). *La bombe informatique*. Paris: Editions Galilée.

PART II.

CRITICAL THINKING FOR A NEW FRAMEWORK

NEW VISIONS FOR A DIFFERENT DESIGN ANTHROPOMETRY

GIULIO CEPPI
Madec, Politecnico di Milano, Italy

1. EVOLUTION OF DESIGN'S PHENOMENOLOGICAL THRESHOLD: NOTES FOR A NEW ANTHROPOMETRY

Industrial Design, as it was commonly defined until some ten years ago, developed to deal with the industrial production of objects[1]. Its phenomenological scale was always that of the object or, in other words, the anthropological measurement of things, with reference to the dimensions and ease of manipulation of the directly visible, as demonstrated first by the British Art and Crafts movement and later by the German *Werkbund*, which achieved conceptual triumph with the Bauhaus in Weimar.

More than 100 years have passed since then and yet, for some, time seems to have stood still and perhaps we all still believe that Design deals exclusively with the visible and definite aspect of objects; with what we can see and touch. This essay seeks not to convince you otherwise but to explain why, in order to deal meaningfully with the visible and tangible side of a product, we must, both in the previous and subsequent stages, address dimensions that are far more complex and therefore equally imperceptible but without which we risk reducing Design to mere cosmetics and styling and losing it as an extremely powerful factor to economic and social development in everyday life.

We will therefore divide our discussion into two stages:
- an initial series of reflections of a historical and evolutionary nature, which aims to provide a better understanding of the extent to which the scope of Design has broadened, moving well beyond the purely anthropometric dimension of the form of things
- a second series of reflections, subdivided into the three phases of inclusion, osmosis and extension (which are also one of the key proposals of this book), which aims to demonstrate how, in the now not-too-distant future, the designer's knowledge and operational toolkit must, indeed, include the scales of the infinitely small and the infinitely large, unless we wish this figure to remain a mere passive observer of the world as a subject matter.

[1.]
Reference should be made to the main texts on industrial design history, the following of which are worthy at least of mention: De Fusco, (1985); Gregotti, (1986); Branzi, (2008).

1.1 The Origin of Industrial Goods and Product Design

In the early 20th Century, new production technologies necessitated new expressive languages. Indeed, Marcel Breuer chairs and Walter Gropius lights are known to all as archetypes of Modern Design; a clear expression of a new, emerging era with, on one hand, machinery and new manufacturing methods and, on the other, man and his material needs. The result was a simple language inspired by the geometric perfection of the cylinder, the square and the sphere, shapes once difficult for craftsmen to produce but which presented no problems for the new industrial machines that came to replace the old ones, with the introduction of new metallic materials, glass and technical fabrics. It is no coincidence that Bauhaus' material production still and always consists of home furnishing objects, tableware, rugs and objects for the workplace, as these are also the product types that serve as preludes in design history; references which I will not bore you by citing or reconstructing here as we are all familiar with them.

It worth noting, however, that the historical perception connoting the term 'Design' was that of the object and its associated anthropometric dimensional scale, starting in the home and the workplace and then quickly graduating to private and public transport. To get an idea of perfect integration between Design and Taylorism, let us bring to mind Henry Ford and his great talent for serial management, verging on the ironic.

1.2 Corporate Culture

Later - although not by much if we consider the work of pioneers such as Peter Behrens - the emerging 19th- century goods design culture expressed an early creative tension and a clear yearning to overcome the dimensional scale of the domestic object, yet with a leap less dimensional than conceptual and phenomenological, encompassing everything that revolves around the object and the identify of its producer: not only the place of manufacture and work but also all the intangible and communicative aspects associated with the design of the brand, packaging and graphics. Let us consider, indeed, the world-famous case of AEG, once again with Peter Behrens and his concern for "management of the visible": lettering, showrooms on Potsdamer Strasse, and even the *Turbinenfabrik* (turbine factory) and employee quarters.

A conceptual leap in which the object was extended and expanded having been defined as its "communicative aura" (Anceschi, 1992) or by its "corporate qualities" (Ceppi, 1996): the foundations were laid for what is, today, defined as the "product system", in which the object's communication, distribution and identity are interwoven with its tangible and physical aspects, thus defining that very hybrid and dynamic *unicum* we call Design.

1.3 From the Spoon to the Town

In post-war Italy, from the '50s and during the boom that followed, Ernesto Nathan Rogers' world-famous slogan interpreted and expressed the fact that Design was increasingly a method through which it was possible to deal with a variety of scales and subjects. Design culture was advancing, expressing different personalities and approaches, in terms of sensitivity and taste.

Often, intersection between the scale of objects, architecture, graphics and art was strong and connotative, particularly if we consider the masters of Italian Design; how else, indeed, can we understand figures like Enzo Mari or Bruno Munari, Marco Zanuso or

Vico Magistretti than by appraising their interdisciplinary approach and fusion of genres and scales? However, while "design culture" had been extended in an increasingly interdisciplinary manner and different dimensions had come into play, we were still in the realms of the commensurable and the visible.

This is confirmed by the great planning orchestration staged in Italy by the company Olivetti - beginning, case in point, in the '50s - in which the graphic and visual scale, the product's dimensions and systemic logic, the architectural conception of space and even the territorial vision (the world-famous zoning plan of 1937 for the Valle d'Aosta Region with the participation of BBPR, Bottoni, Figini and Pollini and coordinated by Adriano Olivetti himself) contribute to demonstrating the great organisational capacity of design culture.

1.4 Radical Design and Utopias

In the '70s, with the arrival of Radical Design in Italy, we witnessed another interesting dimensional and conceptual shift through the well-known work of groups such as UFO, Archizoom and Superstudio. Indeed, Design, metaphysical object and pop culture made a strong entrance at the level of architecture and, indeed, of urban planning and definition of the artificial landscape. Design decisively tackled territorial utopia and the issue of artificial landscape in a large-scale expansion with a healthy dose of irony. Take, for example, Archizoom Associati's *No-stop City* (1970) in which residential diagrams of microclimatised, artificially-lit systems lend themselves to infinite and continuous aggregations, like a great, undifferentiated shelf in a modern supermarket.

However, the exhibition that most celebrated Design's right to completely rethink the *status quo* and define new behavioural and housing models (to intervene, in other words, not only in forms and languages but also in lifestyle, use of space and the dimension of imagery) was undoubtedly *Italy: the new domestic landscape* (1974), an important exhibition held at MoMA, curated by Emilio Ambasz. Here, Design recreated different housing and lifestyle models, from cars to domestic living, expressing strong social and political values, as was typical of the '70s: interiors were defined and treated, in fact, with the emotional power of a landscape, as the title itself metaphorically suggests.

Design showed its social colours in a decisive and provocative way, going well beyond the dimensional facts, playing on intellectual provocation and adding imagery and symbolism to the mix.

1.5 Primary Design and Material Design

The '80s and '90s saw other elements of growth and expansion in Design's phenomenological complexity, with primary design exploring the field of the subjective and the non-visible, contesting the primacy of geometric form's purity and visibility through its theoretical manifestos (Trini Castelli and Petrillo). Smells, tactile sensations and dimensions difficult for Design to describe clearly and competently required new design languages, new tools for observation and analysis of reality, and places like the Montefibre Design Centre and the Fiat Qualistic Centre testified to the new attention paid by Design and Industry to sensory perception and the growing importance of succeeding in visioning and controlling otherwise fleeting parameters, such as smell or sound, which are fundamental in the automotive industry, for example.

As early as 1972, Clino Trini Castelli, through his "Diagramma dolce di Gretel", had been exploring the qualistic dimensions of the home of the sister of the great philosopher, Wittgenstein, defining a visual representation of the primary design in which an emotional and perceptive study claimed a place alongside the compositional one. The subject of sensoriality and perception of qualities other than form alone - including such things as in-depth colour analysis at semantic, perceptive and cultural level - leads us inevitably to the issue of Material Design, celebrated in an exemplary manner by Ezio Manzini (1986) in his world-famous text, *La materia dell'invenzione*, and subsequently tackled by research centres such as that of the Domus Academy (Ceppi, 2014), and the *Istituto Europeo di Design* through the specific subject of Bionics by Carmelo Di Bartolo. To work with the mechanical and chemical properties of materials is to engage directly with the world of chemistry and the sciences connected to the study of polymers and metals, while the more "superficial" work associated with what is known as CFM (Colours, Finishes and Materials) involves all those aspects of encoding and metric referencing that often go beyond the strictly geometric and often purely formal culture of the designer.

1.6 Visioning, Strategic Design and Business Design
In the early 2000s, Design also developed a desire to expressly reach forward in time and anticipate the future, overcoming what could have been perceived as pure Modernity and become an expression of the *Zeitgeist* to come. Indeed, while this phenomenon had already dawned in the '50s, albeit only in the automotive industry through America's the so-called dream cars, a seductive fantasy of what the car of the future would be like, *Vision of the Future* (1996), a project conceived and coordinated by Stefano Marzano of Philips Design, was the first instance of a company systematically and transversally tackling the subject of a future scenario. This project was, indeed, a great media and planning operation consisting of a book, a video and an exhibition, which, based on the research and innovation content shared with Philips Research Lab, narrated possible product and service solutions, arranged on four broad levels: 'personal space', 'domestic environment', 'urban space', and 'on the move'. It was an initial formal declaration of the way in which new technology can generate continuity, connecting the apparently separate and distinct levels and scales on which we humans exist, with our devices and relative connections, moving nomadically through a variety of artificial proxemics.
This systemic approach, capable of integrating technology, consumerism, behaviours and markets within a single vision in which the company is perceived in an interdisciplinary manner as a living organism, was then adopted by new design disciplines as well as associated formative products, as demonstrated by Masters courses such as Strategic Design (Politecnico di Milano, 1998, prof. E. Manzini) and Business Design (Domus Academy, 2005, prof. G. Ceppi). It is, in fact, the dimension of time, not only of spatial and systemic expansion, that comes into play, espousing Design and temporal forecasting and generating a design approach that is not tied to the idea of precise intervention but is fluid and ongoing, increasingly horizontal and open to the dynamics and turbulence of consumerism and the markets.

1.7 Sustainable Design and Service Design

Inevitably, the environmental question, too, required visioning of previously unknown variables and introduction of scales other the purely anthropometric one. Indeed, the planetary and global question necessitated not only a different awareness but also attention to numerics and flows not previously considered and ecosystems of which we are unavoidably part but of which we had lost understanding and awareness. The concepts of life cycle analysis, first, and of traceability (slow food) demanded a circular, as opposed to linear, view of things, slowly shifting from 'green economy' to 'blue economy': previously imperceptible and unknown dimensions became considerable and visible, further extending the perception of the product to its entire life cycle; to a dimension closely tied to Ecology, Biology and Chemistry, well beyond appearance of things.

'Before' and 'after' were added to 'during', with visioning of energy, consumption and pollutants becoming additional levels of visioning and control now necessary for definition of a 'good design'.

The product was increasingly decentralised while the service component became absolutely fundamental, through new, alternative and sustainable-use cultures such as sharing, carpooling and peer-to-peer, virtuous sustainability practices requiring common values and inclusion of new measurements; indeed, the carbon footprint became an absolute measure of value and judgement.

2. DESIGN IN THE AGE OF THE PHENOMENOLOGICALLY 'OTHER'

It therefore seems clear, today, that the intervention threshold of Design is no longer that of the mere object or the scale of the product or furnishing, and yet, leafing through design magazines, it seems that the world got stuck, at least in phenomenological terms, in the '30s and that we are still talking only about yet another infinitesimal variation in form and style.

Obviously, this is not the case. In my opinion, a great early embodiment of the new, enlarged product anthropometry can be found in that marvellous exercise, *Powers of Ten*, a video produced by Ray and Charles Eames using what were then the prototypical methods of an incipient digital culture but already well capable of demonstrating the complex relationship between all things and the value of scales not previously achieved by the traditional contribution of Design. Indeed, in 1974, IBM (then an undisputed leader in the emerging ICT sector), commissioned two American designers to create a trade fair stand. The result, rather than a classical solution set up *in situ*, was a very powerful and elegant visual story, created, at that time, using digital proto-techniques that were pioneering at narrative level, highly rational and precise on one hand but equally visionary and emotionally charged on the other. A journey by powers of ten from the infinitely large to the infinitely small, starting from the familiar scene of a picnic by Lake Michigan, Chicago.

Interestingly, the two extremes coincide in indefinite grey (today, after just 40 years of scientific research, we have come a long way) yet, in the infinite expanse of the cosmos, as in the unexplored energies of the atom, man is, when all is said and done, the threshold and the scale capable connecting and relating such distant extremes. Today, we have gone further thanks to the techno-sciences, yet it is not merely a question of metaphorically visioning or becoming aware of explorations and theories that, at times, appear abstract or speculatively more philosophical than scientific but, rather, of concretely understanding how much Design at such thresholds is, in fact, a decisive and active factor capable of bringing real information and therefore decision and concrete change to processes that used to seem distant but are destined to alter our immediate future more than we think.

To demonstrate this, I will cite the thinking of Giuseppe Testa, director the European Institute of Oncology, who speaks in his book, *Geni a Nudo* (entitled *Naked Genes* in the English translation) about the importance of the 'molecular gaze', that is of appreciating the extent to which the distinguishing feature of the sciences of life, as much as or perhaps even more than their ability to redesign bodies and organisms, is their ability to render visible things, pieces of life, that we were previously unable to see. Today, our gaze no longer penetrates merely the organs but the genes as well: to understand life is to alter it, and knowledge becomes action. These are crucial steps for Design: the knowledge economy is rising to operational levels that make it possible to intervene in life in hitherto unthinkable ways, presenting a challenge that Design must embrace in which the concepts of individual and community, interior and exterior, natural and artificial are profoundly and irreversibly changing.

2.1 The Infinitely Small: Managing the Design of New Life Forms

As previously mentioned, Design has, since the end of the '80s, been exploring the potential of new polymeric materials, thus perceiving that a design threshold can exist in the field of that which is not visible or directly manipulable but which necessitates a meaningful and active dialogue with specialist disciplines such as, for example, Chemistry and Physics. Simply by moving backward in the production chain and starting from the laboratory as opposed to the factory, it is possible to produce innovative, tailored products which, in turn, permit the production of highly innovative industrial objects and products capable of previously unseen performances.

Today, we have achieved scales of intervention far more sophisticated than the one Chemistry and Physic permitted then, as the bio-sciences - Genetics and Molecular Biology - now enable us to intervene in living materials and to alter that which nature intended for them in drastic and unexpected ways.

The relationship between wisdom, knowledge and information guided Medicine for centuries. Today, however, it seems to have undergone further change: 'seeing inside', as in the famous anatomical theatre of Padua, has advanced from the threshold of the organs to that of the genes, and these have now become the object of industrial warehouses and patents. One need only think of Anne Wojcicki, ex-wife of Sergey Brin, founder of Google, who took all 30 billion dollars she had in the bank and invested them in low-cost and DIY genetic tests through her company 23andMe. Is this not ground linked to design? Pharmaceutical companies, hospitals and governments are interested in Genomics; this is not a case of hypochondriac patients but the future of the concept of health and medicine.

In his extraordinary lecture during our seminars, Giuseppe Testa, ranging from Heraclitus to Elliot, Fellini to Foucault and Wordsworth to Walmart, in a highly effective crossing of disciplines, stressed that the study of the reactions of genes to the environment (Epigenetics) has only been possible thanks to digitalisation, and also that the importance of Personal Genomics will, in the future, be linked to the ability, provided by digital networks, not only to calculate but also to connect.

However, Design must enter the research laboratories and have the courage and modesty to move away from the scale of the visible and enter that of 'visioning', where the materials and forms of tomorrow have the potential to form. Testa himself symbolically expressed that we are, today, witnessing a move from the forms of life to the life of forms. In this scenario, responsible and ethical innovation is a crucial issue for scientists, just as the issues of ecology and the environment have been for designers for the last two decades.

2.2 Not Swallowing Technology but Giving a Technical Body to the Enzyme System
Roberto Cingolani, head of the Genoa *Istituto Italiano di Tecnologia* (IIT), impressed upon us during his talks that we can, today, talk about Atomic Design, i.e. design at atomic level. Nature, with its great talent for combining a finite number of elements, still has much to teach us if we seek to take back possession of those concepts of metabolism and circularity of which our prevailing haste and linearity have caused us to lose sight over the last two centuries of desperate industrialisation. Design, too, as we have previously highlighted, has become aware of this, at least since the '90s. Hydrotropism and thermotropism are the life cycles we must appropriate in the near future by investigating materials and their living processes and studying their vital principles. By understanding the vital value of our antibodies, we can ensure a different life for the planet and improve our ways of life.

As early as the '90s, the writings of Derrick de Kerckhove and Paul Virilio were inciting us to 'swallow technology', metaphorically anticipating the way in which nanotechnologies and the micronisation of the digital would, in this way, invade our bodies with sensors, probes and chips. Today, indeed, we understand the workings not only of medical imaging but also of Genomics itself. In truth, it is not only about swallowing what we used to believe it was sufficient to 'tame', but about understanding that our saliva is technology and replicating it artificially on different scales. What we must now accomplish is an epistemological revolution. Indeed, it is no longer technology that is infinitesimally reducing in size to increasingly micronic levels and entering our inner body parts; rather, it is nature that is rising to become the dominant technological model, on a large scale of reference. This is the step to be aspired to in the future. An example of 'ability to see' could be the way in which we all, today, perceive mobile phones as extensions of our individual bodies, connecting us to the global network: tomorrow, by looking at things differently, they could become elements of an encircling vital environment for each person and object, perhaps altering our concept of privacy or even of individual identity, as has already occurred on a genetic scale. Perhaps a kind of 'always on' democracy awaits us, capable of redefining the concept of individual liberty and common good. These are matters that rightfully concern Design.

2.3 The Ability to See the Invisible. Activating and Sharing the "Molecular Gaze"
I believe it is necessary to stress, once again, that the different scales we are talking about must first be approached on a cultural and cognitive level, by adopting a different

viewpoint and by sensitising our perceptual habits to something other; to different worlds and dimensions. It is not a case acquiring superpowers but of entering into the logic of what we have already defined, thanks to Giuseppe Testa, as the 'molecular gaze' or, in other words, the ability to vision that which is as yet unknown. To vision is, already, to intervene, and to bring into focus is to define. This is no different from what happens in the practice of storytelling and stage-setting, where, to define a user type, to create a given mood using the senses, is to define the design values in which we must then intervene operationally. These are operations known to Design at methodological level, and we must now apply them, without hesitation, in a scientific context, flanked, of course, by those with specific expertise in the field and sector.

This is the convergence we aspire to; the quantum leap Design must make if it is not to be left on the sidelines of the transforming world and its materials.

Perhaps a book like that of Penny Le Couteur and Jay Burreson, curiously entitled *Napoleon's Buttons*, can help us to better understand the importance of this argument, albeit retrospectively. Indeed, in their foreword, the two chemists and authors of this book write: "The idea that momentous events may depend on something as small as a molecule – a group of two or more atoms held together in a definite arrangement - offers a novel approach to understanding the growth of human civilization. A change as small as the position of a bond - the link between atoms in a molecule - can lead to enormous differences in properties of a substance and in turn can influence the course of history." In fact, the examples given of Napoleon's army's buttons during the Russian campaign, more so than that of the surrender of Manhattan by the Dutch due to their beliefs about nutmeg, are truly enlightening.

2.4 Copying Nature. Natural Architectures and Biomimetics

Once again taking up the theories of Roberto Cingolani, we should merely restrict ourselves, in future, to copying nature and moving towards the molecular intelligence of its architectures; 'thigmotropism' means generating objects like living organisms and considering Biomimetics, Biometrics and Bionics as models for generating future goods and products, including ourselves. It means re-educating our production models - so that plasticity may take lessons from cellulose - together with our models of consumption and learning to save water by filtering it like nanotechnological mussels. The singular designer Maurizio Montalti provided a practical demonstration, through his 'growing labs', that fungi are powerful recyclers; organic elements suspended between life (air) and death (earth), between fruit (fungus) and plant (mycelium), in a sophisticated and open equation between Engineering, Biology and Design. Decay can coincide with forms of detoxification, generating dirty-clean cycles that are relatively efficient in performance terms, and biodegradation of plastics needs to evolve in that direction. The designer goes back to being an alchemist (whereas the alchemist was, by definition, a designer) in his role of catalyst, straddling Mechanics, Digital Chemistry and Biotechnology. In all these experiments, time becomes a key variable of Design, not only due to the conceptual question of cyclicity but also due to its capacity to involve, on occasion, long time periods, while giving these a value; a deep meaning that we had lost the ability to recognise. We have grown up with the aesthetics of 'expiry'; that things are perfect and then, suddenly, no longer consumable: let us prepare ourselves

for a world that is imperfect yet durable due to recognised and happily perishable qualities, in which the long wavelength of time may reabsorb the errors of the linear and superficial culture that our industrial model has followed, blindly and inevitably, for the last 200 years.
Perhaps we will learn to recognise that there is more intelligence and future in a *Panerochaete chrysosporium* than in a Kevlar composite material; perhaps a chitinous, porous, fibrous future, very different from the anodyne, shiny, glazed imagery that we have always been served up, awaits us.
An interesting response describing the new terminal of the Chinese airport of Shenzhen as 'manta ray', a stingray that breathes, changes its form, unfurls and undergoes variations, transforming into the body of a bird" comes, unexpectedly, in my view, from Massimiliano Fuksas. A continuous honeycomb skin, consisting of metal and glass panels of various sizes, which deforms, acquiring rigidity and flexibility at the same time: 'The idea came to me in New York when a friend gave me a gift wrapped in unusual, honeycomb-style paper. I liked it so much that I threw down the box and started to think about how that three-dimensional geometry could be used in large-scale architectural applications, in such a way as to uniformly resolve the volume as though it had no pillars or supports but was supported by itself: by its skin'. A game of scales; a change of viewpoint, which dimensionally shifts one structure into another, as in a rather pop *transfert*. Is this, too, not a "molecular gaze", perhaps? A disembodied concept of Design must be overcome in favour of one capable of restoring a vision of the body and the "flesh of the world" (as Maurice Merleau-Ponty called it): we should have learned this by now after 20 years of the Slow Food Movement in Italy (and now beyond).

2.5 Skin as an Energy Issue: Perceiving our Physiological Frontier

Chris Bangle, during his talks, spoke to us of 'artifactual elegance', namely the need to combine thinking with elegance, not in an abstract or theoretical way but by observing, once again, the wisdom of certain solutions that exist in nature and having the ability to see their cultural potential. The theme of 'copying' returns once more but, here, the term is synonymous with the 'ability to see', not with blindly imitating; indeed, it means descending to the phenomenological threshold of the world not only as it appears but as it behaves.
For Design, it is, then, about knowing how to combine the two dimensions and integrating the 'how' with the 'why'. This is not merely a question of aesthetics and form but also profoundly of ethics and management, while requiring awareness at non-immediate levels that shift our 'ability to see' to unusual scales and viewpoints.
A design like that of iCub's 'sensitive skin', endowed with a neuromorphic system of thousands of receptors, is entirely focused on energy consumption and generating sensitive systems that are energy-efficient. Indeed, the current computational systems consume millions of times more energy than ours, and the brain, a highly efficient plantoid, remains a thus-far inimitable model. The energy issue is central to skins and sensors. Robotics, nanoparticles and intelligent polymers dance together, producing wonders and lending themselves to new uses: is this not Design? At the IIT, there are 900 scientists from 37 different countries, but perhaps still few designers.

Returning, however, to our old body and the potential of Biometrics, we know that our skin is becoming part of a far more sophisticated scenario. Indeed, according to Fujitsu, palm vein scanning enables biometric scanners to identify people and prevent fraud: a light with a near-infrared wavelength links us to a chip and then to the vastness of the web, verifying, first, that blood is actually flowing through the veins in order to avoid the risk of false positives. The application is already at the advanced test phase in 15 shops in Lund, Sweden. The digital keys of the future are the heartbeat, smell, vein pattern, iris and even the biometric signal produced by the mind, which will, perhaps, in future, enable us to drive a car in an exclusive man-machine relationship.

The consequent possibility of digital cloning is, evidently, close at hand; indeed, the video *New dimensions in testimony*, available on YouTube, demonstrates how, using 6,666 LEDs and three high-speed cameras, it is possible to record an entire figure and reproduce it in a totally hyperrealistic way, capturing every detail. Where does our skin physically end and what, as the psychologist Didier Anzieu explains so well in his wonderful book, is the boundary between us and the world? What certainty of this boundary now remains? Is it up to the anxieties of the 'Quantified Self' - using bracelets and other 'apps' to meet the need to constantly and obsessively measure what is happening inside and outside ourselves - to resolve the problem?

For some, such as Matteo Lai of Empatica, it is, instead, a matter of 'affective computing'; detecting psychological and physical stress to help us and re-educate us for a better life.

I believe these are questions which Design must seek to answer and in which it can be an essential, responsible, active and significant element, not merely an aesthetic practice in the service of marketing, choosing the best fashionable colour and the most attractive packaging.

2.6 Third Syndrome: Equilibrium as a Perennial Dynamic

We know that, in recent years, there has been an explosion of 'thirdness' or, in other words, the tendency to add a third, balancing factor to two pre-existing factors. Indeed, Gilles Clement spoke to us of the 'Third Landscape', and the artist Michelangelo Pistoletto spoke of the 'Third Paradise'. In fact, he went as far as to model an interesting generative theory called 'Trinamics', based on the power of the number three.

Today, however, Pistoletto no longer defines himself as an artist but, rather, as an "arti-vist", i.e. an artist/activist who makes things happen; ambassador of a future who, through his work, generates aggregation tools in order to impact on society, thus altering it. Is this not a perfect definition of a designer, he who has always interfaced between production and consumption, ownership and enjoyment, past and future? The subject of the future is no longer, for either an artist or a designer, a search for one's own identity, now dissolved into a babel of languages in which everything can make sense if justified and motivated. Identity used to be, as before Pistoletto's famous metaphorical mirrors, a kind of "continual transition" in which identity was the search for one's own rationality and the search for oneself coincided with maximum freedom. A work of art therefore explained how things were; a narcissistic and autotelic exercise, indeed. Today, the mirror has value if it becomes an aggregation; no longer a solitary reflection but an inclusion of subjects in a third space, be it real or virtual, but, rather, in which the *status quo* is altered, probably including the mirror itself. Ethics, responsibility and sustainability are the new values of Art. Provoking, in a search for new forms of equilibrium. Denouncing, to pave the way for new opportunities. Indeed, Art, like Design, exists if it transforms; if it produces reality. Otherwise it is a limited exercise, befitting a fashionable gallery or a glossy magazine; we may all talk about it on the social networks, post it and google it, yet nothing will actually happen.

3. THE OUTSIDE RETURNING INSIDE: THE LONG DANCE OF EXTERIORISATION AND INTERIORISATION
As early as the mid-'90s, as part of *The Solid Side. Il lato solido in un mondo che cambia*, a project coordinated by Ezio Manzini with various players such as Michele De Lucchi, Stefano Marzano and Clino Trini Castelli, among others, we designed, together with Luca Gafforio, what we defined as 'Skinware Commodities' or, in other words, transcutaneous neo-goods which displayed accurate and targeted information and 'on-air' messages on the skin itself. The implicit message behind the design provocation was that we should use our own bodies, like an organic and dynamic tattoo, to take back ownership of that which, phenomenologically, belongs to another, virtual and fluid, world.

In a way, it was a precursor to what we define here as the 'outside returning inside' in an inversion of that apparently linear, subtle and progressive process effectively described by the anthropologist André Leroi-Gourhan in his study, *Gesture and Speech*. Indeed, it is as though our history of techno-cultural civilisation corresponded to a kind of continual exteriorisation; an infinite expulsion of objects, prosthetics and devices for functions that our bodies have not succeeded in satisfactorily resolving, as has, on the other hand, occurred for other animal species which have made a specialty of their greatest assets as strategies for evolution and adaptation. Walking upright, the erect posture, specialisation of the cranium and consequent freeing-up of the hands have enabled us to transfer to tools, equipment and instruments and therefore more sophisticated methods of controlling nature and the exterior world to the point of bending them (almost) to our will.

Perhaps we are now at the beginning of a parallel phenomenon: a change of direction, having externalised everything it is possible to externalise, such as our brain into a PC and our social relationships into a network. We are at the point of (parallel and synchronous) inversion while, from another point, the line, continues to grow, albeit asymptotically, toward extreme levels, as I believe we have already described in the previous point. The 'Internet of Things' proposes a scenario made up of billions of sensors; a network of networks, in which systems and things hold a dialogue, seemingly coming at us and creating an infinite prosthetic belt through which we must literally learn to navigate in order to survive in a universe of bits. Atoms and bits fused together, for precision. This is a scenario also discussed by Stefano Marzano who, for years, sustained through his intriguing designs of future visions that we should put an end to those boxes (be they black or white) that we still call computers and aspire, instead, to a reality made up of diffuse and distributed intelligence, 'Ubiquitous Computing' and intelligent objects to be considered rather like modern butlers or house-stewards capable of serving us with silent courtesy. His designs in partnership with Felice Rossi and Cappellini are theoretical manifestos of the way in which exteriorisation can be replicated, poetically, through a kind of re-interiorisation of technology, passing via archetypes almost immutable over time, like an armchair or a bookcase.

3.1 Nature Re-tamed: The City-World as an Antidote

The scale in question, whose inversion we need to capture, is not only symbolically our bodies but also our very environment, *in toto*. For example, we are witnessing the countryside - expelled from and placed in contrast to the man-made urban context for the last two centuries - returning victoriously to the city; indeed, 'vegetecture' proposes a different concept of greenery in the city which is highly integrated and destined to alter our lifestyles. We are not talking about making wine on a terrace in Manhattan, as some affable billionaires amuse themselves by doing, but defining new behavioural and production models. From Emilio Ambasz, via Peter Brooks and Andrea Branzi, to the vertical agriculture guru Dickson Despommier, durability and care, production and consumption, living beings and life cycles are the concepts that are changing the design of the future supercities. Educational and therapeutic gardens, sensory gardens, and night gardens for Alzheimer's sufferers ('healing gardens') are just a few of the areas with which 'Evidence-based Design' is experimenting, an approach that seeks to generate economic and psychological benefits for city dwellers.

Marc Augé, during his talk, clearly explained to us the dichotomy between the world-city (real and local) and the city-world (virtual and fluid). The latter is the equivalent of the world as a spectacular landscape; the illusion of a single, great, connected planet, as seen from the window of an orbiting space capsule. Ubiquity and instantaneousness dominate our culture, ensuring dilution without boundaries and expansion without places, turning us all into globalised tourists. These decentralised cities, consisting of interiors in which the global wins over the local and of standardised exteriors, must be colonised once again, from the bottom up, with the culture of services and Design, where Architecture and Urban Planning have failed miserably. Today, Car2Go is worth more than an Area Zoning Plan, an educational garden teaches more than a course in the classroom, and a green festival moves more people than a union meeting.

I have learned more from the Reggio Children schools and nurseries, where I have been fortunate enough to spend time over the last 20 years for various projects (most recently the water and energy workshop in Ligonchio), than from the Institute of Architects' training courses, compulsory though they may be.

3.2. Tackling the Infinitely Large: Learning to Trace

According to the writings of Giovanni Bignami (head of the Italian Space Agency) in L'Espresso magazine (2017, the 7th May), we are truly 'children of the stars', as the well-known Italian song, *Figli delle stelle*, goes. The Integral satellite has shown us that iron is born and compounded in the sky due to an explosion of gamma ray energy (which only iron can emit), just like carbon. These are the two building blocks of life, something the red of our blood should remind us of. Thermonuclear fusion, which occurs every day in the sun, teaches us that we are made from the same elements, from the carbon of our tissue to the calcium of our bones, both also produced by stars. In other words, the infinitely large is part of us and we must not, therefore, consider it so distant: first and foremost, we should get to know it, intimately; some would say paradoxically.

It is not, therefore, a question of foreseeing life on Mars but perhaps of pondering intriguing visions such as those of *The Grand Design*, proposed by that astrophysicist of rare narrative and theoretical talent, Stephen Hawking: "Between the initial state of a system and our later measurement of its properties, those properties evolve in some way, which physicists call the system's history", as the author Richard Feynman defines it. These 'histories' implicitly demonstrate that no concept of reality in fact exists independently of descriptions or theories. In short, to use a term dear to Design, 'storytelling' is fundamental to Science, too, and taking possession of it is not far removed from what is occurring today in Design, particularly when we are researching a new path towards innovation.

Indeed, Stefano Marzano clearly illustrated for us that to innovate is to sail against the wind, something which takes skilful and patient tacking, to using a sailing metaphor, but also that it always requires a 'roadmap'; a vision of the 'other', open to things to come, since attempting to establish new archetypes and deciding who has the authority to do so is no trifling matter.

We have learned from the Slow Food Movement that, on the subject of dietary archetypes, traceability (i.e. the concept of a chain) is an important principle for assessment of any agri-food product. To trace means to connect the here and now (what we find on our plates) with something bigger and more complex: the locality that produced it, the animal and plant species that made it possible, and the processes and treatments that gave it its defining characteristics. In a way, it means creating a kind of product cosmology; defining its material universe of relationships and connections, and therefore establishing its value and quality. This awareness of the relationships that connect material and man, natural and artificial universes, is therefore no less fascinating and complex that what the Integral satellite is telling us. Understanding the connections, connecting the local with the global, the part with the whole, is a fundamental challenge for mankind and, today, with religions and mythologies no longer seeming sufficient, Design, too, has an important role to play in the process of accounting for the meaning and value of things.

3.4 Designing from the Bottom Up: Creative Communities

The society of 'prosumers' envisioned by futurologists and pseudo-economists like Jeremy Rifkin and Chris Anderson anticipates millions of people in a position to produce and share energy, materialising objects created in 3D. Indeed, there are those who speak of a third industrial revolution (to add another element to the hunger for 'thirdness' described above), in which business will no longer be B2B or B2C but what is already defined as H2H (Human to Human). They speak of cutting production costs, to the benefit of everyone, by creating novel social networks in continual transmigration. Perhaps, though, there is a less extravagant and simplistic path, as Raymundo Sesma has shown us through his patient and charismatic artwork: once again, as in the case of Michelangelo Pistoletto, Art comes alive and interacts with life in a direct and profound manner.

To cite what Piero della Francesca was already saying more than 50 years ago, maintaining that, in a painting, the relationship between two cypress trees is more important than the depiction of a single cypress, Art explains that connection and involvement are, today, a very powerful design tool; a kind of Buddhist vision of a world made up of infinite and invisible relationships which, when activated, can transform our perception of what we simplistically define as 'real'.

It is not, however, the utopia of education for all or, worse still, of the ubiquity of knowledge, as googling on the internet seems bent on ensuring, which may be perceived as a form of democracy or, worse still, participation. The Internet remains a means, not an end; the illusion of freedom, not a guarantee of being free. As Marc Augé reminded us, God, still to this day, reserves the right to meddle in human lives; a kind of right of intervention, and one which design culture, too, must reserve, albeit with greater modesty: the ability to provoke, with due respect and appropriate intelligence.

[Follows commentary via images.]

References

Anceschi, G. (1992). *L'oggetto della raffigurazione*, Milan: Etas.

Anzieu, D. (1992). *L'epidermide nomade e la pelle psichica*, Milan: Raffaello Cortina Editore.

Branzi, A. (2008). *Design italiano*, Milan: Electa.

Ceppi, G. (1996). *Progettare le corporate qualities. Linea Grafica,* 305.

Ceppi, G. (2012). *Design Storytelling*, Milan: Fausto Lupetti.

Ceppi, G. (2014). Il design dei materiali in italia. il contributo del centro ricerche domus academy 1990-1998. AIS/Design. Storia e ricerche, 4:0408.

Cingolani, R. (2014). *Il mondo è piccolo come un'arancia*. Milan: Il Saggiatore.

De Fusco, R. (1985). *Storia del design*, Milan: Laterza.

Gonzalez-Crussì, F. (2014). *Organi Vitali*. Milan: Adelphi.

Gregotti, V. (1986). *Il disegno del prodotto industriale*, Milan: Electa.

Hawking, S. (2011). *Il grande disegno*. Milan: Mondadori.

Leroi-Gourhan, L. (1982). *Il gesto e la parola*. Turin: Einaudi.

Le Breton, D. (2007). *Il sapore del mondo*. Milan: Raffaello Cortina Editore.

Le Couteur, P., & Burreson, J. (2008). *I bottoni di Napoleone*, Milan: Tea.

Manzini, E. (1986). *La materia dell'invenzione*, Milan: Arcadia.

Nowotny, H., & Testa, G. (2012). *Geni a nudo*. Milan: Codice Edizioni.

Evolution of design's phenomenological threshold
THE ORIGIN OF INDUSTRIAL GOODS AND PRODUCT DESIGN

Sedia B32 "Cesca" - 1928
Marcel Breuer

Sedia Wassily - 1925
Marcel Breuer

Sedia F51 - 1920
Walter Gropius

Servizio the TAC - 1969
Walter Gropius

Bauhaus Manifesto - 1923

Total Tool

Evolution of design's phenomenological threshold
CORPORATE CULTURE

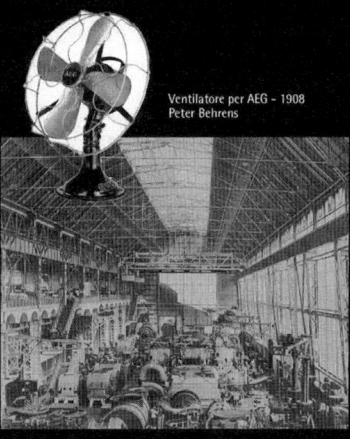

Ventilatore per AEG - 1908
Peter Behrens

Fabbrica turbine AEG - 1909
Peter Behrens

Manifesto lampadine elettriche - 1910
Peter Behrens

Fabbrica turbine AEG - 1909
Peter Behrens

Francobollo AEG - 1910
Peter Behrens

Total Tool

Evolution of design's phenomenological threshold
FROM THE SPOON TO THE TOWN

Superleggera - 1955
Aldo Rossi

Seggiolina - 1964
Marco Zanuso

Centro assistenza
paesi africani - 1971/74
Marco Zanuso

Nuvola rossa - 1977
Vico Magistretti

Appartamenti Piazzale Aquileia - 1964/5
Vico Magistretti

Piano regolatore
Valle d'Aosta 1936/7
Gruppo coordinato da Adriano Olivetti

Total Tool

Evolution of design's phenomenological threshold
RADICAL DESIGN AND UTOPIAS

Conica - 1980/83
Aldo Rossi

No-stop City - 1970
Archizoom

Mobile and Flexible Environment Module - 1972
Ettore Sottsass
Italy: the new domestic landscape (1974)

Mini-kitchen - 1963
Joe Colombo
Italy: the new domestic landscape (1974)

La casa calda - 1984
Andrea Branzi

Total Tool

Evolution of design's phenomenological threshold
PRIMARY DESIGN AND MATERIAL DESIGN

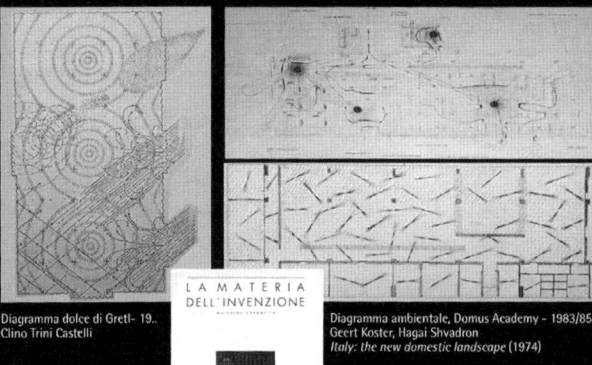

Diagramma dolce di Gretl- 19..
Clino Trini Castelli

Diagramma ambientale, Domus Academy – 1983/85
Geert Koster, Hagai Shvadron
Italy: the new domestic landscape (1974)

Neolite.
La metamorfosi delle plastiche- 1991
di E. Manzini

La materia dell'invenzione – Arcada,1986
Ezio Manzini

Total Tool

Evolution of design's phenomenological threshold
VISIONING, STRATEGIC DESIGN AND BUSINESS DESIGN

Vision of the future – 1996
Stefano Marzano (Philips)

WIRED – 1996

Business Design – 2005
Domus Academy, Giulio Ceppi

Design strategico – 1998
Politecnico di Milano
Ezio Manzini

Total Tool

Evolution of design's phenomenological threshold

SUSTAINABLE DESIGN AND SERVICE DESIGN

Slow Food - 1986
Carlo Petrini

Carbon footprint

Car Sharing - Bike Sharing

Blue Economy - 2010
Pauli Gunter

Total Tool

DESIGN IN THE AGE OF THE PHENOMENOLOGICALLY 'OTHER'

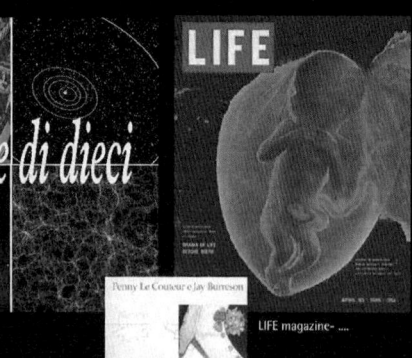

Potenze di 10 - 1968
Ray e Charles Eames

LIFE magazine- ....

Knowledge Economy - 1992
Peter Drucker

I bottoni di Napoleone - 2006
Penny Le Couteur, Jay Burreson

Total Tool

Inclusion

THE INFINITELY SMALL: MANAGING THE DESIGN OF NEW LIFE FORMS

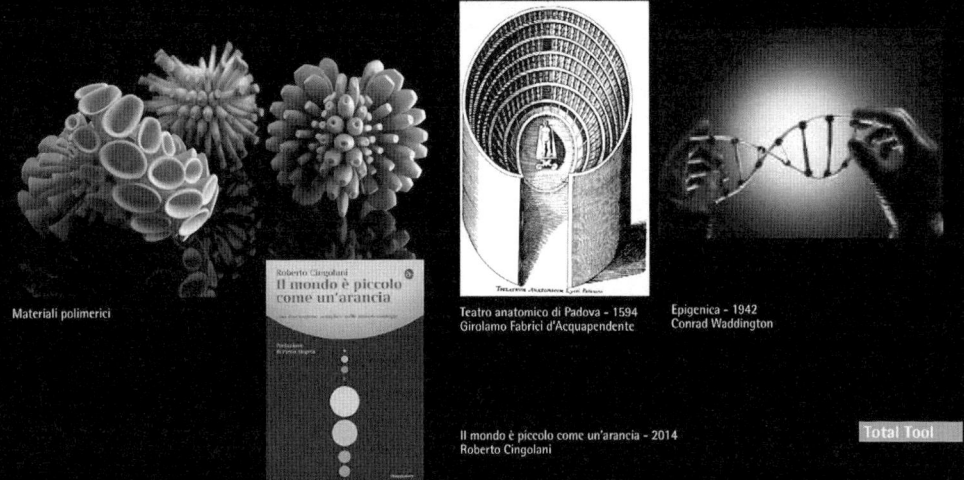

Materiali polimerici

Teatro anatomico di Padova - 1594
Girolamo Fabrici d'Acquapendente

Epigenica - 1942
Conrad Waddington

Il mondo è piccolo come un'arancia - 2014
Roberto Cingolani

Total Tool

Inclusion

NOT SWALLOWING TECHNOLOGY BUT GIVING A TECHNICAL BODY TO THE ENZYME SYSTEM

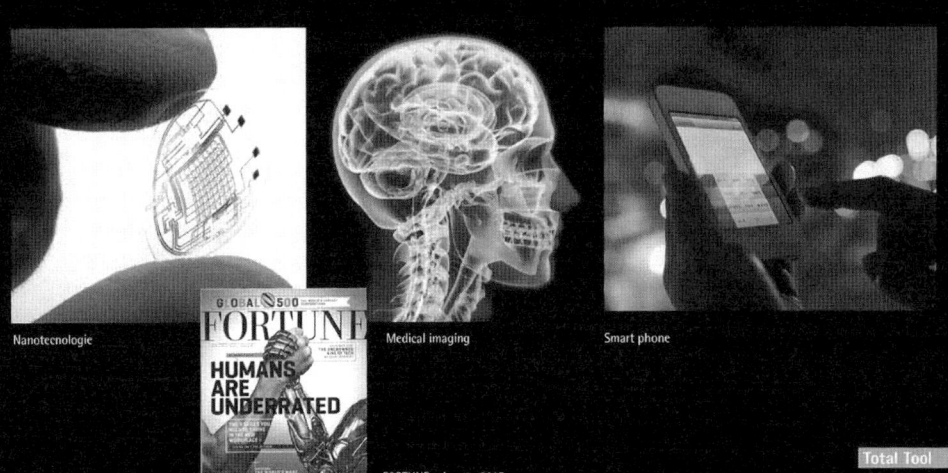

Nanotecnologie

Medical imaging

Smart phone

FORTUNE - Agosto 2015

Total Tool

Inclusion
THE ABILITY TO SEE THE INVISIBLE. ACTIVATING AND SHARING THE 'MOLECULAR GAZE'

Storytelling Sguardo molecolare Personas, Utenti

Geni a nudo. Ripensare l'uomo nel XXI secolo - 2012
Giuseppe Testa

`Total Tool`

Osmosis
COPYING NATURE. NATURAL ARCHITECTURES AND BIOMIMETICS

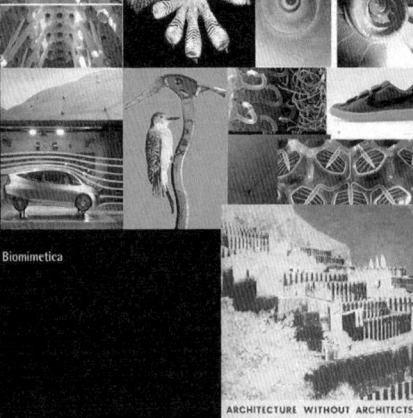

Biomimetica

Growing Labs
Maurizio Montalti

Aeroporto Shenzhen-Bao'an - 2013
Massimiliano e Doriana Fuksas

Architecture without Architects - 1964
Rudofsky Bernard

`Total Tool`

Osmosis

SKIN AS AN ENERGY ISSUE: PERCEIVING OUR PHYSIOLOGICAL FRONTIER

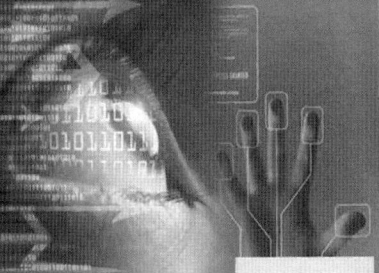

Biometrica

iCub - 2009
IIT Genova

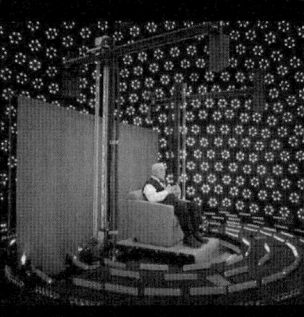

"New dimensions in testimony" - 2013
USC + ICT

L'epidermide nomade e
la pelle psichica - 1992
Didier Anzieu

Total Tool

Osmosis

THIRD SYNDROME: EQUILIBRIUM AS A PERENNIAL DYNAMIC

The third man - 1949
Carol Reed

Il terzo paradiso - 2003/12
Michelangelo Pistoletto

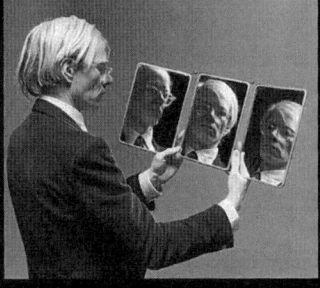

Specchio, aggregazione, inclusione di soggetti

Manifesto del terzo paesaggio - 2005
Gilles Clément

Total Tool

Extension

THE OUTSIDE RETURNING INSIDE: THE LONG DANCE OF EXTERIORISATION AND INTERIORISATION

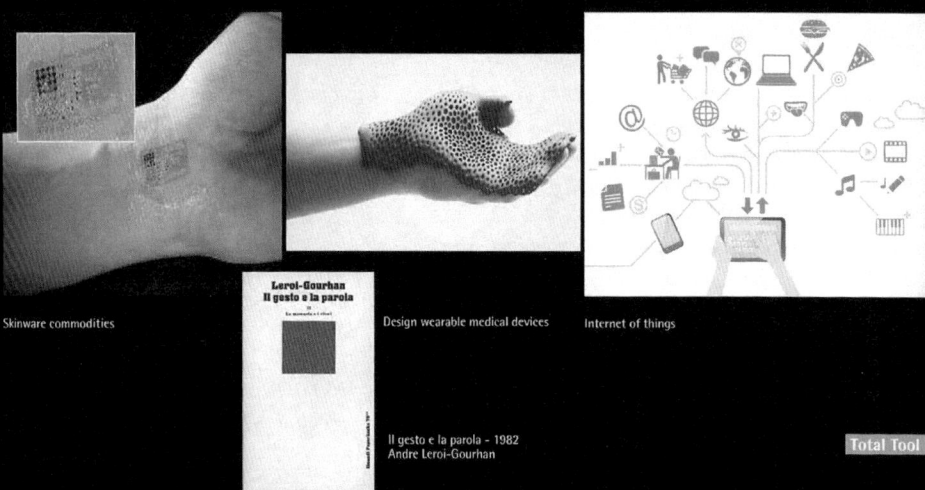

Skinware commodities

Design wearable medical devices

Internet of things

Il gesto e la parola - 1982
Andre Leroi-Gourhan

Total Tool

Extention

NATURE RE-TAMED: THE CITY-WORLD AS AN ANTIDOTE

Healing garden

Casa del Ritiro Spirituale - 2004
Emilio Ambasz

With Something for the Donkeys and Farmhouse Teas and Children Welcome - .....
Peter Brooks

Il gesto e la parola - 1982
Andre Leroi-Gourhan

Total Tool

Extension

TACKLING THE INFINITELY LARGE: LEARNING TO TRACE

Satellite *Integral* - 2002

Storytelling-Storyboard

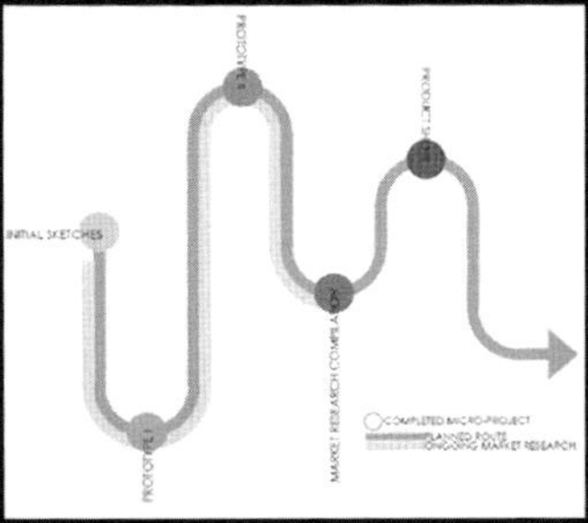

Roadmap

Il gesto e la parola - 1982
Andre Leroi-Gourhan

Total Tool

Extension

DESIGNING FROM THE BOTTOM UP: CREATIVE COMMUNITIES

TOWARD THE BIOCENTRIC ERA. OBSERVING DESIGN HYBRIDIZATION

CHIARA LECCE

Madec, Politecnico di Milano, Italy

Go, take your lessons from nature, that's where our future lies
Leonardo da Vinci

Biocentric ethics merged at the end of the 1980s as environmental ethics that "extend the status of moral object from human beings to all living things in nature" (Yu & Lei 2009, p.422) starting from Albert Schweitzer's ethic of reverence for life (1987), Peter Singer's ethics of animal liberation (1990) and Paul Taylor's ethics of *bioegalitarianism* (1986). This new environmental ethic approach called for a new prospective: "nature does not exist simply to be used or consumed by humans, but that humans are simply one species amongst many" (Bari, 1995), that translated means for *Homo Sapiens* is not superior to other species on Earth.

Biocentric ethics are founded on three main principles: "(1) All living things have an instinct to resist the process of increasing entropy, for keeping their own organization, maintaining their own survival and the wholeness of life; (2) Maintaining their own survival is a central aim for all living things, that is an intrinsic value and 'good'; (3) Although different living things have their own ways of organization and survival (different ways of self-organization and maintaining survival with different organisms), their values are intrinsically the same, and therefore they should have equal rights in morality, which means they should be given moral acknowledgement, concern, and protection". (Yu & Lei 2009, p.422-423)

More recently another exponent of the *biocentric* vision has been Robert Lanza, an American doctor and scientist involved into several researches of the medical use of human embryonic stem cells.

He affirms in his book *Biocentrism: How Life and Consciousness are the Keys to Understanding the Universe followed* (2009) that biology will be the most important above all the other sciences.

"Our theories of the physical world don't work and will not ever work as long as they will start to consider life and consciousness. Life and Consciousness, instead of tardive and secondary products appeared after billions of years of inanimate physical processes, are absolutely essential for our comprehension of the universe. We call this prospective *biocentrism*." (Lanza 2015, pp.13-14)

1. DESIGN HYBRIDIZATION: BRIDGING DIFFERENT DISCIPLES
Definition: in biology *hybridization* is the process of combining different varieties of organisms to create a hybrid. *In situ hybridization* (ISH) is a type of hybridization that uses a labeled complementary DNA, RNA or modified nucleic acids strand (i.e., probe) to localize a specific DNA or RNA sequence in a portion or section of tissue (*in situ*), or, if the tissue is small enough (e.g., plant seeds, Drosophila embryos), in the entire tissue (whole mount ISH), in cells, and in circulating tumor cells (CTCs). *In situ* hybridization is a crucial step for understanding the organization, regulation, and function of genes.[1]
The term hybridization could be applied today to many more fields (per se an explicative phenomenon of disciplines cross-fertilization): chemistry, linguistics, engineering and automotive, globalization theories and media.
I chose biology and genetics definitions because they are closer in their meanings regarding the process of design hybridization that I would like to analyze during this dissertation.
The design profession faced during the last two decades several political, social, economic and environmental conflicts. Accordingly, contemporary design panorama is seeking a great expansion toward many different directions in a continuous transfusion with other disciplines (from economy to computer sciences, from sociology to psychology, from engineering to biomedics), in a sort of attempt to resolve all the problems of the world. Borrowing the words of Paola Antonelli: "In the past 20 years, design has branched out in many new directions that have galvanized young practitioners, sparked business models, and set the worldwide education system on academic fire. There are many different ways in which one can be a designer today, working for instance on interaction, interfaces, the Web, visualization, typefaces, socially-minded infrastructures, 5D spaces, sustainability, games, critical scenarios, and yes, even products and furniture. In the coming decades, the rest of the world will catch up and design will be embraced as a methodology and philosophy by politicians, scientists and economists who are willing to have a human, holistic and constructive perspective on the world." (2011, p.110)
And this is exactly what is already happening. Theories of design-driven-innovation are invading all the managerial master courses, environmental topics are at the basis of all design schools, social design and service design are inseparable from product design strategies.
Complexity is the keyword and designers are asked to use their skills to create a better future. Here emerges the concept of Hybrid Design: "It's a progressive notion about the multi-dimensional craft of 'doing things', as well as a reflection on the interconnectedness of all kinds of design within the economic and commercial fabric of society. [...] Hybrid design breaks these professional silos and asks the design team to be aware,

1.
https://en.wikipedia.org/wiki/In_situ_hybridization

intelligent and reactive to an eco-system of experts surrounding the design process. Hybrid designers re-design, re-think and, in time, reflect on their work in progressive new ways. Over time the work coming out of a hybrid design team is of a better quality, better suited to a complex physical/non-physical world and better positioned to weather the tests of time, society, and culture." (Amit, 2010)

Approaching design hybridization concepts the words of the American designer and educator Victor Papanek from his book *Design for the Real World* (1985) appear more actual than ever: "Design must become an innovative, highly creative, cross- disciplinary tool responsive to the true needs of men. It must be more research-oriented, and we must stop defiling the earth itself with poorly-designed objects and structures. [...] It is at the border between different techniques or disciplines that most new discoveries are made, most action is inaugurated. It is when two differing areas of knowledge are forcefully brought in contact with one another that, as we have seen in a previous chapter on bionics, a new science may come into being".

For Papanek the interdependence of various disciplines could be explained quoting a story by Buckminster Fuller: "In the last decade, two important papers were presented to learned societies, one on anthropology and the other on biology. And both these researchers were working completely independently. But it happened by chance that I saw both papers. The biological one was looking into all the biological species that have become extinct. The anthropological one was looking into all the human tribes that had become extinct. Both researchers were trying to find a commonality of causes for extinction. Both of them found the same cause independently - extinction is a consequence of over-specialization. As you get more and more over-specialized, you inbreed specialization. It's organic. As you do, you outbreed general adaptability. So here we have the warning that specialization is a way to extinction, and our whole society is thus organized." (1985)

Papanek in conclusion proposes three main tenets: "(1) Design of products and environments, on or off earth, must be accomplished through interdisciplinary teams; (2) biology, bionics, and related fields offer the greatest area for creative new insight by the designer; (3) design of a single product unrelated to its sociological, psychological, cityscape surroundings, is no longer possible or desirable. Therefore, the designer must find analogues, using not only bionics but biological systems design approaches culled from the fields of ecology and ethology."

And it will be exactly this last statement our core argument of the following paragraphs.

2. LESSONS FROM NATURE: FROM BIONICS TO BIO-DESIGN

Since the begging of human history, nature has always been a permanent lesson. Humans apprehended from nature rhythms, balances and energy and above all its uncontainable power of survival and defense.

The words of the famous German architect and engineer Frei Otto will help introducing the relationship between man and nature and its interpretation: "The idea that the objects of nature, especially objects of living nature, can be taken as a model has always moved many people. It is the base of many religions: 'God created the world with it plants, animals and man. What God created cannot be doubted, it is perfect. And the most perfect being – man – is made in his image'. It is certain that most biological

objects can look back on a long evolutionary history, and because of this long selection of the less effective they may be considered optimized. Many biological objects, such as a few unicellular organisms (bacteria, radiolarians, algae, etc.) had reached a high degree of perfection already in their early developmental history, changing hardly or not at all ever since.
They have lived for millions of years, distinguishing themselves mainly by a very high adaptability to their surroundings and in particular to their hosts. [...] Many of the organisms living today are extremely complex systems." (Otto 1999, p.48)
And he also introduces *bionics* as the new science, the aim of which is "finally" to learn from nature. As an illuminated man of the XX Century he understands the importance of the nature lesson but he also admits the limits of his century to deeply comprehend all the mysteries of nature:
"[...] History best proves how little the model nature was exploited. No biological objects served as a model of the great inventions of humanity: the hammer, knife, wheel, the woven cloth, the bow, ect. One reason may reside the fact that living nature is so very complex that is difficult to really get to know it. [...] A technician observing living nature just cannot grasp living objects which die so quickly, are so sensitive, so complex and both so inimitable and strange. A biologist looking at technology sees how imperfect technical activity is." (Otto 1999, p.48)
Still at the end of the XX Century the referees to the natural world were associated more from a mechanical or formal point of view because of a lack of adequate technologies that were on their way coming in few years. And this necessity is also well explained in this passage by Giampiero Bosoni with Francesca Picchi: "The history of design in all its expressions - technological, typological, morphological and, we would add, poetic, has in diverse ways and with diverse results measured its strength against this unavoidable model of reference. [...] We do however care deeply about the fact that nature's lesson cannot be read only as an evolutionary mechanical, and in that sense exclusively technical principle. Nor perhaps, can it be learnt even from the mathematical-geometric point of view only, according to an abstract and purely synthetic conception of the system of rules. Nor, still less, can it be properly understood by tritely imitating its forms, emulating the form of organic growth with curved lines, soft masses or rounded figures." (1999, p.55)

2.1 Bionics and biomimetic

The term *bionic* was coined by Jack E. Steele in 1958, from the word *bion* (Ancient Greek: βίος), meaning "unit of life" and the suffix -*ic*, which means "like", so together "like life". Bionics focused on extracting shapes, structures and function from nature to create close copies of natural artefacts: it is oriented to a technological transfer between the natural world (seen as a highly optimized and efficient system) and human manufactures. Typical examples of bionic applications in engineering are the hulls of boats imitating the thick skin of dolphins, sonar, radar, and medical ultrasound imaging imitating animal echolocation or also the dirt- and water-repellent paint (coating) from the observation that the surface of the leaf of the lotus plant (the lotus effect).
Copying from nature is also the method defined by Roberto Cingolani in this book (pp.) with the term *atomic design*: nature as the most perfect system is the only model to pursue.

The name *biomimetics* was coined by the American biophysicist Otto Schmitt in the 1950s. A clear view of the potentialities of biomimetic approaches is reported by Jean-Pierre Ternaux, French neurobiologist, honorary director of CNRS, in his contribute to the book *The industry of nature: another approach to ecology*: "The use of biomimetics methodologies is above all concerned with scientific research, the aim of which is to facilitate the study of natural systems by laboratory reproduction of phenomena hidden in the complexity of real organisms. [...] Apart from its interest in advancing knowledge relating to living organisms, the biomimetics approach offers the possibility of developing innovative therapeutic applications. In this context, current research concerns the development of nano probes and intelligent nano vectors and control systems to allow drugs ti be transported within an organism to target specific cells. Application of biomimetics methodology also concern improvement of diagnostic techniques by the development of high-quality molecular imaging, the production of biomaterials for restorative surgery, the development of new virtual-reality and augmented-reality technologies in the service of both the health and pathology of mankind. Just like mimicry which concerns not only the individuals of a species but also their relatives and their social organizations, biomimetics today takes inspiration from knowledge acquired in the domain of ecosystem organization, or more generally from the functioning of living organisms, to better integrate organizational principles and human technologies in the societal domain." (2012, p.17)

Lastly Ternaux introduces synthetic biology discipline which relies on the same principles of biomimetics: "This relatively new scientific field combines research in the fields of biology with engineering principles, aimed to design and construct (synthesize) new biological forms and systems with controllable functions. In this field, the use of the principles of biological engineering, such as *in vitro* models, standardization, normalization, automation and computer-aided design, all contribute to simpler, faster, less expensive processes to modify living organisms".

2.2 Biomimicry

Since the beginning of the XXI Century progresses in biological research have made great leaps forward. Molecular biology and genetics, "have permitted to unveil logics, principles, languages and codes on which nature is based, from the macro-scale down to the molecular level and beyond. [...] In this regard, inspiration from nature has a much broader significance, oriented towards the transfer of biological principles and logics". (Langella 2010, p.295)

Here we can introduce the term biomimicry, from the Greek *bios*, "life", and *mimesis*, "imitation". Biomimicry's main theorist is the American naturalist Janine M. Benyus who proposes for the first time this new approach in 1997 with her book titled *Biomimicry: Innovation Inspired by Nature*.

"The biomimics are discovering what works in the natural world, and more important, what lasts. After 3.8 billion years of research and development, failures are fossils, and what surrounds us is the secret to survival. The more our world looks and functions like this natural world, the more likely we are to be accepted on this home that is ours, but not ours alone."[2]

Benyus explains that she started to understand the "real lessons" from nature studying wildlife habitats and behaviors: "Where ecology meets agriculture, medicine, materials science, energy, computing, and commerce, they are learning that there is more to discover than to invent.
[...] Our fragmentary knowledge of biology is doubling every five years, growing like a pointillist painting to a recognizable whole. Equally unprecedented is the intensity of our gaze: new scopes and satellites allow us to witness nature's patterns from the intercellular to the interstellar. We can probe a buttercup with the eyes of a mite, ride the electron shuttle of photosynthesis, feel the shiver of a neuron in thought, or watch in color as a star is born. We can see, more clearly than ever before, how nature works her miracles. [...] We realize that all our inventions have already appeared in nature in a more elegant form and at a lot less cost to the planet. Our most clever architectural struts and beams are already featured in lily pads and bamboo stems. Our central heating and air-conditioning are bested by the termite tower's steady 86 degrees F. Our most stealthy radar is hard of hearing compared to the bat's multifrequency transmission. And our new smart materials can't hold a candle to the dolphin's skin or the butterfly's proboscis. Even the wheel, which we always took to be a uniquely human creation, has been found in the tiny rotary motor that propels the flagellum of the world's most ancient bacteria."[3]
The chapters of the book define all the fields of action of the Biomimicry Institute founded in 2006 by Janine Benyus and Bryony Schwan:
"Nature runs on sunlight.
Nature uses only the energy it needs.
Nature fits form to function.
Nature recycles everything.
Nature rewards cooperation.
Nature banks on diversity.
Nature demands local expertise.
Nature curbs excesses from within."[4]
One of the most interesting and useful contribution of the Biomimicry Institute is the platform called "AskNature" created and launched in 2008 as "an audacious project" seeking to catalog and present all the biological knowledge collected in the form of an open source database for designers, architects, entrepreneurs and all "that non-biologist innovators across disciplines".[5] Their database is structured in two main sections: Biological Strategies and Inspired Ideas (design solutions to human challenges, inspired by biological strategies).
Biomimicry approach has led significant applications developed by laboratories, companies, designers from all around the world.
This approach involves both the natural and the artificial world generating new projects that could include a wide range of typologies: products, materials, energy production, systems, architectures and even organizational and systemic strategies for sustainable solutions (Bistagnino, 2009).
Bio-inspired design is a form of adaptation to the complexity of contemporary age triggering a contamination process between natural sciences, bioengineering, design and many other different disciplines. (Salvia, Rognoli, Levi 2009) This process has been

defined also Hybrid Design by Carla Langella from the Second University of Naples and from the Sapienza University of Rome. (Langella 2007)
"The definition 'hybrid design' derives from the field of tissue engineering: hybrid biomedical tissues are obtained in laboratory using techniques reproducing biological processes of cell growth. When implanted in damaged areas, hybrid tissues, albeit designed, processed and characterized following an engineering approach, are perfectly compatible with the biological organism, responding, therefore, to specific medical needs. In the same conceptual way, hybrid design prefigures artefacts with characteristics which are intermediate between nature and technology, and whose genesis and evolution can in itself be defined as hybrid. Hybrid design, therefore, widely refers to technological transfer from fields with high scientific and technological content. This transfer constitutes not only a procedural and methodological approach, but also an inexhaustible source of available instruments and technologies, through which bio-inspired concepts are translated into hybrid products." (Langella & Santulli 2010, p.295)
The short following selectin of projects have been all generated by biological systems. They seek to a bio-inspired design which could be spread through several disciplines and different kind of applications.

Materials
BioFriend Anti-Microbial (2011) is a molecular filtration technology able to trap and kill microbes. Developed by Filligent (HK) Limited company this material takes form by a biomimicry solution: the sites on human cells which are able to destroys microbes surfaces (viruses) and cell walls (bacteria).
The *insecta* animal class is above all one of the most successful biomimicry subject. The *NanoSphere®* (2011) self-cleaning, and dirt-water-repellent fabric finish by Schoeller Technologies AG company, originated by observing butterflies' wings nano-scale surface structures able to repel water and dirt. Spiber[6] is a Japanese research company focused on proteins based material that should substitute plastics, metals and other synthetic materials in the future. In 2015 Spiber collaborated with The North Face company for the application of a spider fibroin-based protein material that is considerate on of the toughest material on earth, to create a new prototype of outwear clothes called "Moon Parka". Other insects-inspired projects are "fog-catching" materials. Oxford biologist Andrew Parker and Chris Lawrence of QinetiQ studied tenebrionid (*Stenocara*) beetles in the barren Namibian Desert. The shell of these insects has

2.
http://www.nytimes.com/books/first/b/benyus-biomimicry.html
3.
Ibid.
4.
http://biomimicry.org/
5.
https://asknature.org/
6.
https://www.spiber.jp/en

a corrugated surface texture, these hydrophilic bumps surrounded by hydrophobic troughs allowed the beetles to collect water from fog in order to survive in arid environments. "The materials inspired by these beetles have many potential uses, including tent coverings and roof tiles for collecting water from fog in regions lacking access to fresh water".[7]

From the sea world, the Marine Biological Laboratory Woods Hole (Massachusetts) in 2012 started to study *Optical metamaterials*[8] from squids' skin which is able to perceive light and change color. Over squid's skin occurs a sort of signal processing, then the chromatophores and iridophores produce a colored pattern in response. "Mimicking this complex system could have interesting applications i.e. for camouflage and displays".[9]

Ecosystems

A smart biomimicry approach observes not only chemical and physical proprieties of living creatures but also how their ecosystems work. *Biolytix* water filter (2013) by Biolytix Water Australia Pty Ltd, is a water filtering system cleans without chemicals converting raw sewage, wastewater, and food waste into high quality irrigation water on site taking inspiration from forest wastes decomposing processes. "The system removes solid wastes from wastewater and then selected worms, beetles, and microscopic organisms convert the waste into structured humus, which acts as a filter to turn the waste into garden irrigation."[10]

7.
https://asknature.org/idea/fog-catching-materials/#.WOKQ-BKLSi4

8.
"Metamaterials" are a new area of research, enabled by advances in sophisticated sensing and materials production. Unlike other materials, these are inhomogeneous composites design to exhibit unique behaviors.

9.
https://asknature.org/idea/optical-metamaterials/#.WOKOGxKLSi4

10.
http://www.biolytix.com/

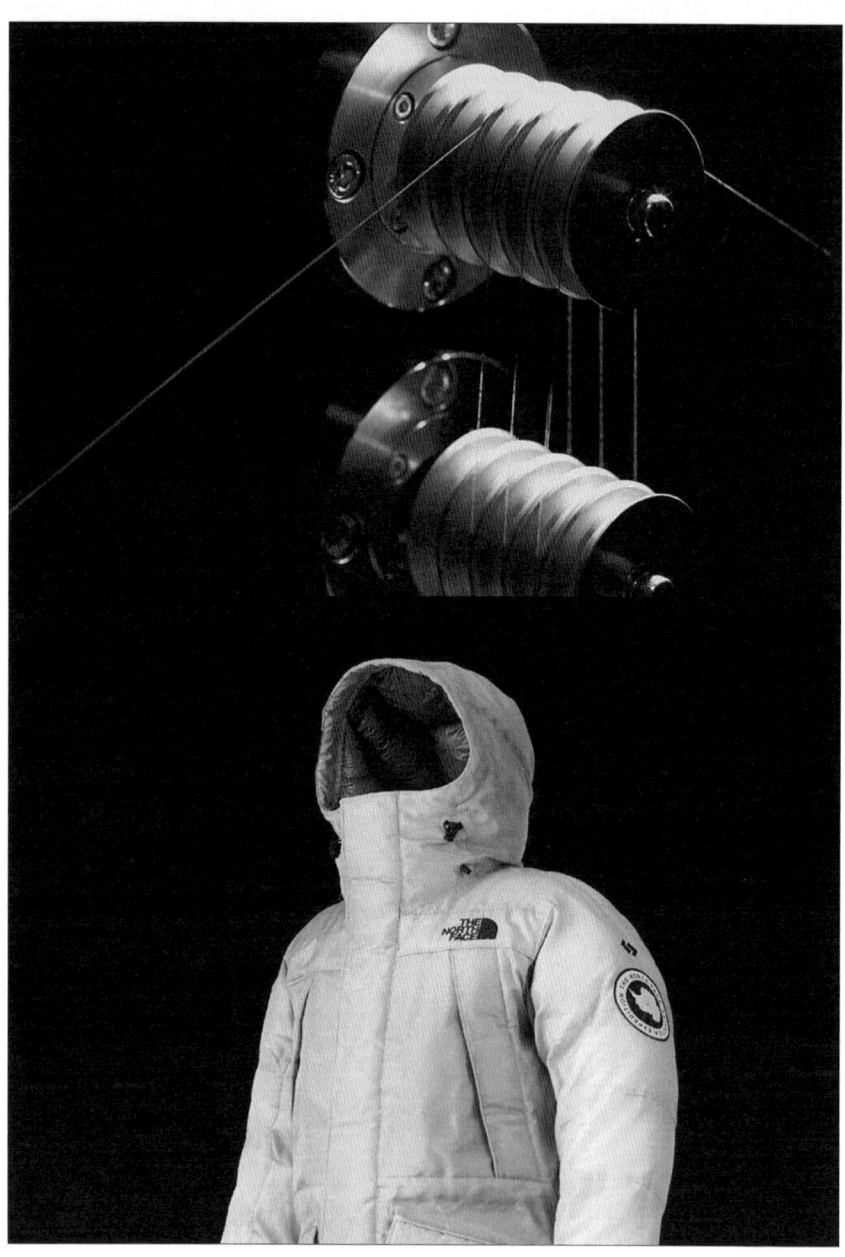

Fig.1 Spiber company collaborated with The North Face for the application of a spider fibroin-based protein material to develop a new prototype of outwear clothing called "Moon Parka".

Architecture
Of course architecture is one of the most involved discipline into the process of biomimicry. *Homeostatic facade* by Decker Yeadon (2014) is a self-shading system for buildings. Decker Yeadon designed a homeostasis biological based system made of a double-skin glass façade which reacts to the internal temperature of a building balancing it between through ribbons of an elastomer wrapped over a flexible polymer core.[11] *Plyskin* by Lindey Cafsia from the Royal Academy of Art in The Hague (Netherlands) is an insulation material inspired by polar bear skin. "The polar bear can survive extremely low temperatures thanks to its efficient fur and skin which consists of three layers: the outer layer, the fur, consists of guard hair and shorter dense underfur, this fur is slightly translucent, so the heat of the sun can penetrate to the second layer, namely the skin. This skin is black, which absorbs the heat of the sun. The third layer is blubber and functions as insulation".[12] So she designed *Plyskin* insulation membrane which consists of three layers: the outer hairy layer is made from recyclable polyamide; the second layer has a honeycomb structure, which makes the panel tense, absorbing part of the heat radiation; the third is a black, empty layer filled with heat absorbing material, functioning as heat buffer. These two layers are made from PET and biobased PLA.

Energy
Clean energy is maybe at the base of a bio-inspired philosophy. Among many experimentation *Polymer-cellulosic batteries* (2009) by the Angstrom Laboratory (an interdisciplinary laboratory at Uppsala University, Sweden) uses algae cellulosic cell structure. Cellulosic batteries incorporate conductive polymers to form essentially a paper battery, but many problems with a limited surface area for charge storage pushed Angstrom Lab to look at *Cladophora* algae. It is a common type of green algae from the Baltic area and around the world. Algal cellulose actually has a fundamentally different nanostructure from terrestrial plant cellulose forming an excellent template for surface modification to include a conducting polymer.[13]

3. BIO-DESIGN
The exploration of these new stimulating and fascinating sciences chased the cross-pollinations among different disciplines brought to the definition of Bio-design. "Bio-designers are turning their attention to familiar organisms like plants and animals. In some cases, they examine the less accessible world of bacteria and cells, while in others they purpose the creation of new living systems by directly manipulating DNA. This endeavor requires collaboration and interaction among different disciplines and is carried out chiefly in groups, raising implication that collide with our deepest belief. […] Their work encouraged and celebrated in a few centers of "irradiation" of these new

11.
https://materia.nl/article/homeostatic-facade-system/
12.
https://materia.nl/article/plyskin-insulation-polar-bear-skin/
13.
https://asknature.org/idea/polymer-cellulosic-batteries/#.WOKORhKLSi4

Fig.2 *Homeostatic facade* by Decker Yeadon, 2014.

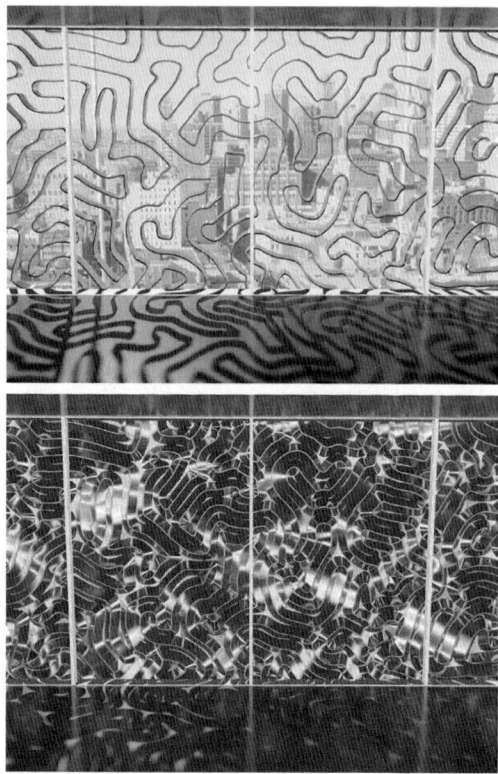

Fig.3 *Plyskin* insulation membrane by Lindey Cafsia from the Royal Academy of Art in The Hague (Netherlands), 2016.

ideas, such as the Design Interaction program at the Royal College of Art in London, the Science Gallery at Trinity College in Dublin, and the Paris-based Le Laboratorie gallery and research center founded in 2007 by David Edwards, a professor of Biomedical Engineering at Harvard, which helped designer Mathieu Lehanneur propel himself in this realm. Many of them were featured in MoMA's 2008 exhibition *Design and the Elastic Mind* and are now part of the museum's collection." (Antonelli 2011, p.110)
Paola Antonelli recalls a series of works that stay at the border between art, design probes and biology. As in the case of Mathieu Lehanneur who developed between 2006 and 2007 a series of projects observing natural cycles reformulating new functions. He used plants and fishes to conceive new domestic appliances: the *O Oxygen Generator*, was a domestic breathing machine using oxygen-producing Spirulina platensis algae; the *Bel-Air* organic-filtering system, also based on plants - Gerbera, Philodendron, Spathiphyllum and Chlorophytum are among the most effective - absorbing the toxins emitted by manufactured goods in our domestic environment; and *Local River*, a home storage unit for live freshwater fish aquaponically combined with a mini vegetable. Among many others, Antonelli mentions also the work of Tobie Kerridge, Nikki Stott and Ian Thompson's with *Biojewellery* (2007). They created rings out of the bone cells of a couple, which are extracted, harvested onto a bioactive ceramic scaffold and then combined with precious metals to finish the rings. (Antonelli 2011, p.113)
Neri Oxman is also another leading proponent of this hybrid scenario presenting her renewed opera at the intersection between computational design, digital fabrication, materials science and synthetic biology.

3.1 Bioutilization: Form follows Nature

While biomimicry focuses on the translation of biological principles into human-made technology, *bioutilization* directly leverages organisms or biological materials. Bioutilization is particularly useful in cases where replicating complex biological machinery or processes in our own technologies is unsuccessful, too time-intensive, or too difficult to be cost-effective. This last examples have been all generated using living organism as their material of construction.
Ecovative Design utilizes fungal mycelium (see also the exemplary work of Maurizio Montalti and his Officina Corpuscoli, p.71) to produce environmentally-friendly materials. Using mycelia and agricultural waste like corn stalks, the company has developed a compostable material as an alternative to plastic foam and other environmentally-destructive synthetics. Mycelia are grown on the agricultural waste, forming a matrix that binds the fibers together and results in a solid mass. The mass is heat treated to stop the growing process and then is ready for use. Ecovative's materials are used in packaging, building materials, and even surfboards, all with comparable performance and cost-competitiveness to other materials on the market.
Protein by Tessa Silva turns milk into bioplastic. "The project explores methods of processing the protein that can be extracted from milk, called casein, as a natural alternative to oil-based polymers. Casein can be turned into a bioplastic and it can be treated as a commercial thermoplastic allowing for compression molding, injection molding, or it can also be hand sculpted for a more crafted and organic results."[14]
That's It (2016) is a biodegradable packaging from algae-based material developed

Fig.4 Mathieu Lehanneur, *Local River home system*, 2008.

by Austeja Platukyte. During her final thesis at the Vilnius Academy of Arts (LT), she followed the Zero Waste philosophy searching for biodegradable packaging material. She chose algae based material, the agar, dissolved in water with calcium carbonate, and a vegetable based emulsifying mix that is used to impregnate the material: the result is a bowl-like packaging material. In 2017 the *Living Colour* bio-design research project born from the collaboration between Laura Luchtman and Ilfa Siebenhaar that explores the possibilities of natural dying with living microorganisms: bacteria. Together with the TextileLab in Amsterdam, they investigated on the idea of "Dancing bacteria": "the optimum growth conditions for bacterial pigments, ways to speed up the growth process and the possibilities of growing bacteria in patterns by subjecting them to sound frequencies. Asking 'What effect do sound frequencies have on the growth of bacterial pigments?' and 'Can we control the process of growing bacterial pigments?' This way we hoped to exclude random growth in order to upscale the bacterial dye process. Growing bacteria as a dye factory can lead to a more sustainable way to color the world."[15]

14.
https://materia.nl/article/protein-turning-milk-bioplastic/
15.
https://issuu.com/kukkadesign/docs/living_colour-ibook

Several multidisciplinary studios are also working on more artistic and future probing projects.
A strong example is the *Regenerative reliquary* (2016) by media artist Amy Karle. She grows bone along a biofriendly 3D printed lattice using medical CAD and human stem cells, using 3D scan data of bones from the California Academy of Science's collection and then rendering the data and applied generative algorithms to create sculptures.
The Austrian Living Studio developed the *Fungi Mutarium* project in collaboration with Utrecht University in order to study a fungi food product able to grown on plastic waste: "growing food on toxic waste".[16]

4. CLOSING THE ISSUE: IS DESIGN A KIND OF MAGIC WAND FOR EVERYTHING?
Concluding this long overview over the world of Bio-suffixed new disciples and their interactions with design professions, I would recall a statement by Paola Antonelli that matches well also my point of view: "Experiments with design are often considered directional or speculative, and designers can indicate new behaviors and unexpected applications, a focus on human life that might at times elude scientists. Although I have shied away from the bombastic declaration that designers can change the world, thanks to these collaborations they just might." (Antonelli 2011, p.113)
Is not possible to pretend from designers to become maximum experts of synthetic biology, robotics, chemistry or materials engineers. And here comes again absolutely pertinent the words of Victor Papanek: "While the designer in any team situation may know far less psychology than the psychologist, far less economics than the economist, and very little about, say, electrical engineering, he will invariably bring a greater understanding of psychology to the design process than that possessed by the electrical engineer. By default, he will be the bridge."
Indeed, designers MIGHT be sensible collectors, attentive observers of both the natural, the artificial and the social world, seeking for open-collaboration, looking to biodiversity in order to avoid sterile self-referencing and waste expanding projects.

16.
http://www.livinstudio.com/fungi-mutarium/

Fig.5 Ecovative Design, Mushroom® Packaging, 2015.

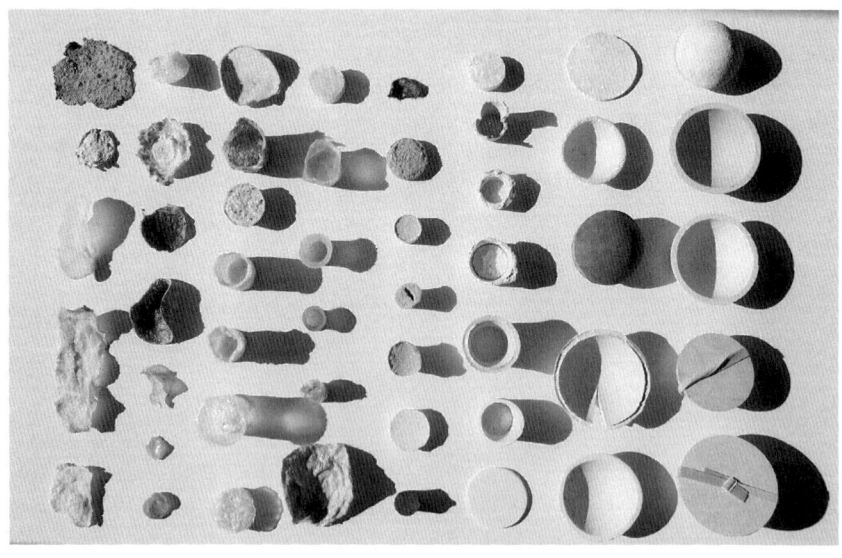

Fig.6 Austeja Platukyte, *That's It*, biodegradable packaging from algae-based material, 2016.

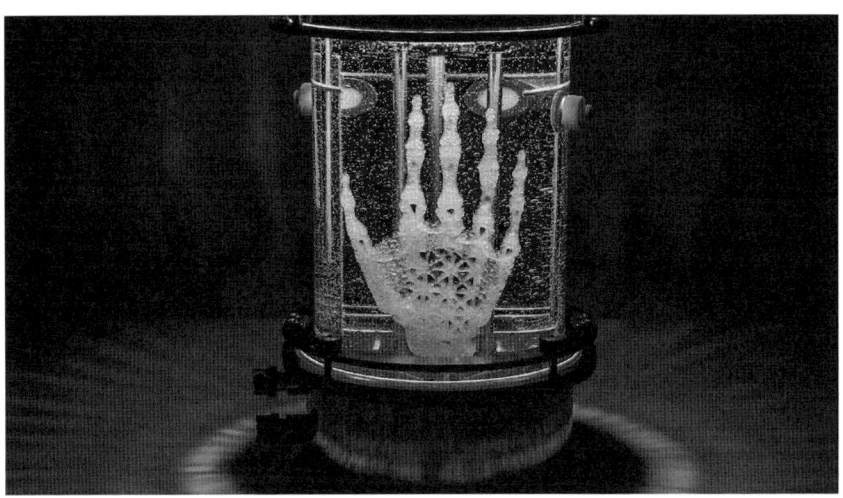

Fig.7 Amy Karle, *Regenerative reliquary*, 2016.

References

Amit, G. (2010). *Beyond Design Thinking: Why Hybrid Design Is the Next New Thing*. Retrieved on 31/03/2017 https://www.fastcompany.com/1656288/beyond-design-thinking-why-hybrid-design-next-new-thing.

Antonelli, P. (2011, November). States of Design: Bio-design. *Domus 952*, pp.110-113.

Bari, J. (1995). Revolutionary Ecology: Biocentrism & Deep Ecology. *Alarm: A Journal of Revolutionary Ecology*.

Benyus, J. (1997). *Biomimicry: Innovation Inspired by Nature*. Harper Perennial.

Bistagnino, L. (2009). *Design sistemico. Progettare la sostenibilità produttiva e ambientale*. Slow Food.

Bosoni, G. & Picchi, F. (1999, September). La Nature, leçon permanent. *Domus 818*, p.55.

Langella, C. (2007). *Hybrid design. Progettare tra tecnologia e natura*. Franco Angeli.

Lei, Y., Yi, M. (2009). 13. Biocentric Ethical Theories. *Environment and Development – Vol. II*. China. p. 422.

Leopold A. (1987). *A Sand County Almanac, and Sketches Here and There*. New York: Oxford University Press.

Otto, F. (1999, September). On nature, the model. *Domus 818*, p.48.

Papanek, V. (1985). *Design for the Real World: Human Ecology and Social Change*. Chicago Review.

Salvia, G., Rognoli, V., Levi, M. (2009). *Il progetto della Natura*. Franco Angeli.

Santulli C. & Langella C. (2010). Hybridisation between technology and biology in design for sustainability. *Sustainable Design, 1*, 3.

Schweitzer A. (1987). *Philosophy of Civilization* (trans. C.T. Campion). Buffalo, N.Y.: Prometheus.

Singer P. (1990). *Animal Liberation*. New York: New York Review of Books.

Taylor P. (1986). *Respect for Nature: A Theory of Environmental Ethics*. Princeton, N.J.: Princeton University Press.

Ternaux J.P. (2012). "Mimicry: a life-saving reflex in species". In *The industry of nature: another approach to ecology*, Frame, pp.15-18.

ETHICS AND DESIGN IN THE 21ST CENTURY

GIAMPIERO BOSONI

MADEC, Design Department, Politecnico di Milano

Reflecting on some of the contributions to the cycle
of conferences headed "Ideas and matter"

During the cycle of conferences dedicated to the topic "Ideas and matter – What will we be made of, and what is the world going to be made of?"[1], the ethical question on conducting researches, designing and, especially, applying or producing the results of research and design, has risen time and again, in more or less open terms. The question has been raised in different meta-design contexts and through different thematic approaches, both as regards phases of theoretical research and as regards applied conditions and processes.
Although we all live in an age which, on various occasions and in several ways, summons all of us to social and cultural, if not actually political, commitment, towards certain issues of an essentially ethical nature, such as, for instance, ecological, health, privacy and cohabitation problems generally, the ethical question is not so frequently tackled in the field of design culture, at least not here in Italy. It was accordingly interesting and stimulating to establish a dialogue, within the School of Design of Politecnico di Milano, through a series of contributions, expression of quite a broad range of viewpoints, which offered us and up-to-date critical interpretation of the ethical issue in the field of design.
Let us start from the first contribution, put forward by the geneticist Giuseppe Testa, director of the Stem Cells Epigenetics at the European Institute of Oncology (and also, inter alia, co-founder of the PhD in fundamentals and ethics of Life Sciences), that introduced us to the extremely delicate topic of decoding the human genome, with all the ethical implications that concern the patentability of parts of our DNA, down to the well-known question marks raised around the debate relating to the notorious "embryonic stem cells". To give an idea of the extent of the problem, Professor Testa reminded us that "as stated in a book by the person who generated Dolly, Ian Wilmut, 'we have entered the era of biological control', and control – as Testa underlines – is of course an essential element of design: when one does certain things and designs them, he would then also like to control them somehow."[2] About this fundamental ethical aspect of control by the designer over the results of his own research, Testa warns us

by recounting to us the extraordinary developments achieved in his project field: "[…] Will we ever be able to do what an egg is capable of doing? Doing it ourselves by "playing" with genes? Practically all the scientists answered in the negative, yet not only is it possible, but from 2006 to 2014, in eight years, we are contemplating the idea of moving from Vuitton to Wal-Mart."[3] As if to say that genetic engineering has already shifted from the sophisticated handmade level of the luxury item to the consumer industrial product. These important statements from Testa remind us of the far-sighted reflections by Siegfried Giedion dating back to nearly seventy years ago, in his enlightening book *Mechanization takes Command* (1948). In a book entirely devoted to the topics, by then still very current, of mechanization, during a period in which design culture was still in its infancy, Giedion decides to devote the last chapter to those which, at that time, were no more than vague post-mechanisation scenarios, especially to an issue still mysterious in those days, nearly futuristic and intelligible to just a few insiders: "genetic engineering". "[…] Around 1930 – Giedion said in 1948 – a new development on the verge of acquiring significant proportions commences. Its peculiar characteristics is to tend towards an epoch that abandons the mechanistic viewpoint. It consists in the intervention inside the organic substance, in the intervention in the structure of animals and plants, in the intervention in nature; man as demiurge. Genetics, a science derived from biology, deals with these interventions. […] In itself, therefore, genetics is nothing really new. It treads the usual path of experience that is traditional to scientific experimentation. This stage lasts long, but what occurs in the epoch of full mechanization may not be compared to previous stages. The interventions in the body go much deeper. What ensues from it is the mutation in the structure of plants and animals at such a speed that, if compared to the earlier one, eliminates the time factor. Size turns gigantic. The revolution underway may only be compared to the transformation that, one century earlier, was undergone by the artisan's work tools, which suddenly took on new forms or metamorphosed into machines. […] More than in any other field, – Giedion goes on to say – in this one a realization is necessary, namely, that precisely out of love of production it no longer perceives everything from a production viewpoint. What is at stake here is not a given quality of iron or steel, a specific type of engine or fridge. What is at stake is in fact the essence of life, something, in other words, that is bequeathed from one generation to the other. […] Which goal are these interferences with the most intimate substance pushing us towards? What are the creatures generated through these systems? How much can production increase by shortening the duration of breeding

1.
Politecnico di Milano, School of Design, October-Dicember 2014.

2.
See "Human in the Molecular Age" by G. Testa in this book, p.18.

3.
Idem, p.18.

and animal life? Where are the boundaries vegetable and animal organisms interpose before these new interferences? [...] One thing is still certain – Giedion concludes –: mechanization must grind to a halt wherever it finds itself before the living substance. A new stance has become necessary. The boundless field of interventions in fertilization, of which today we see but the first developments, places on us a fearsome responsibility. If, instead of devastation and exploitation, we want to attain a genuine control over nature, utmost caution is required: no action by force, but rather the most intimate and in-depth knowledge of organisms should be the starting premise. This enjoins a spiritual conception that decisively gives up the idolatry of production."(Giedion 1948, pp.246-256) A long quotation necessitated by the prestige of the person enunciating it, and given especially that this quite prophetic thought was put forward in a text rightly considered, by many of us historians of design, one of the most important documents on the establishment of design culture at international level.

Still going through the interventions made at the cycle of conferences "Ideas and matter", at another research level we come across the reflections by Roberto Cingolani, scientific director of the Genoa-based Italian Institute of Technology (IIT), who likes labelling himself an "atomic designer", i.e. one specialized in "copying" (as he in fact underlines with a somewhat witty satisfaction), although we might call it, more fittingly, reinterpreting and transposing some essential features of living beings and all the things found in nature into possible applications for the sake of meeting our needs. "The atomic design approach we use is an approach based on the copy – explains Cingolani –, we are the exact opposite of what you seek to develop; we copy from the most evolved system that exists in the universe, namely, the evolutionary system of life that in 3.5 billion years has led architectures to be functional and has established a unique relationship between form, complexity and functions, which is the one enabling us to live. [...] We complain of genetically modified organisms because we think they are dangerous, but in actual fact we are narrow-minded, since the whole of nature is a genetically modified organism, the only difference being that when we modify organisms we do so over two years, whereas nature took a few million years, yet everything is genetically modified organism. This is in fact the weak point of nature, as it takes long to change, it also takes long to adapt, whereas our technologies are changing too rapidly and, therefore, we are often too fast for genetic modification."[4] In that regard, Cingolani seemingly induces us to confront the risk of running too quickly, with little margin for error and verification, in reconstructing in vitro the thousand-year old processes of evolution of natural forms. "The final message – thus Cingolani concludes his intervention – is that we have reached the point that, with great humility, we must copy; that nature has already optimized a series of processes and relationships between form and function, that are so evolved and advanced that only a lunatic might think of improving on it in this moment, hence, as the Japanese used to do in the 50's, we copy and hope that in thirty, forty years' time, our technologies are so reliable and so indestructible that we might then put forward truly novel ones "beyond the evolution", now it is still too early."[5] This theme of the speed of research and the alleged capacity to overstep nature through the creation of new organisms resurfaces in several interventions put forward for the conference cycle. Quite peculiar and significant in respect of the ethical themes of design is also the research presented by Maurizio Montalti who has been engrossed for a period in the study of some micro-organisms, more precisely a kind of fungi, very important inside natural rhythms as regards the cycles of decomposition, of transformation of all kinds of organic and inorganic substrates. Such microorganisms are the great recyclers of the natural world: thanks to their skills, they cover a whole series of activities that enable us as well as the planet to survive. "My designs – Montalti recounts – intend in fact stimulating a degree of tension as regards some possibilities the use of microorganisms might offer to the society we live in, and encouraging the "truly" skilled persons to further this research. In this sense, too, critically stimulating designers and the whole design field on the need to adopt a critical approach to the materials we ourselves daily employ. In order to establish a contact between our disposable culture, the toxicity of plastics and the possibility of having fungi capable of degrading these materials and killing these eternal materials, I decided to focus attention on this globally recognized iconic object that is the famous 'monobloc chair', the garden chair. An extremely cheap object that breaks easily (if you visit a landfill, you will find a huge quantity of these broken chairs), which is why for me this object somehow represents a declaration towards the life cycles of consumer products compared to the immortality of the very materials making up the majority of consumer products."[6] This ethical position, understood as a kind of militant commitment to the detection of critical issues in our consumerist system, though in actual fact in our lifestyles as well, and more generally in our sense of civic responsibility, practiced also with the intention of awakening the attention of designers and of scientific researchers as well, undoubtedly represents an interesting model to ponder in terms of participation and implementation of a valid ethical approach to design.

In support of this position, it might be interesting to go again through some passages of Vilém Flusser's text "Un'etica del disegno industriale?" ("An ethics of industrial design?") (2003), where a reflection is conducted on the fact that the issue of the morality of objects, the moral and political responsibility of the designer, has acquired over the years "a new meaning (or better put, a new urgency)" for at least three reasons: "Firstly, rules are no longer laid down at public level. Although there are still authorities (religious, political, moral) as they existed before, their rules can no longer pretend people's trust; their skill regarding industrial production is doubtful. [...] Their skill is questioned since industrial production has become extremely complex and rules of any kind tend to be misleadingly simple. [...] The only authority that seems to have remained more or less intact is science. Nevertheless, it declares itself committed to an evaluative research, hence it provides technical, rather than moral rules. [...]. Secondly – Flusser again explains – industrial production, inclusive of the design phase, has turned into a complex network drawing information from a number of sources. [...] Accordingly, it has become necessary to work in a team, combining human and artificial components; the results, therefore, cannot be ascribed to a single author. The design phase envisag-

4.
See "Atomic Design and the Artificial Elegance" by R. Cingolani in this book, p.64.
5.
Idem, p.64.
6.
See "The Growing Lab" by M. Montalti in this book, p.71.

es a high degree of division of work and cooperation. Due to this, no person may any longer be deemed responsible for a product." (Flusser 2003, pp.63-64)

In this connection, Flusser forcefully underlines the increasingly more sensitive issue of the ethical question in the shift from the first to the second industrial revolution and, still more, from the second to the third revolution, the so-called post-industrial one: "The lack of moral responsibility that represents the logical consequence of the production process inevitably gives birth to morally reprehensible products, unless we reach an agreement on the type of code of ethics the design activity should uniformly be subjected to. [...] Thirdly, in the past there was a tacit agreement to the effect that the moral responsibility for a product had to be traced solely to the person utilizing it." (Flusser 2003, p.65) Nowadays, Flusser explains, resorting to some paradoxical (but not too much so!) examples, given the ever more frequent use of automated applications and various forms of artificial intelligence, it is increasingly harder to identify the "persona" that makes use of it, and increasingly more complex to define the chain of subjective and objective responsibilities. "In other words – Flusser further writes – if designers do not tackle these issues, a scenario of total lack of responsibility might arise. It is not a novel problem, of course. It has emerged with terrifying clarity in 1945, when a decision had to be taken on who should be deemed liable for the crimes against humanity perpetrated by the Nazis. At the time of the Nuremberg trials, a letter written by a German industrialist to a Nazi officer was found. The industrialist begged forgiveness for having built the gas chambers commissioned to him poorly: instead of killing thousands of persons at once, only hundreds of them would get killed there." (Flusser 2003, p.66) To this dramatic reminiscence by Flusser, it is pregnant to correlate a similar reflection by Giedion drawn from his notes on the topic of mechanization: "[...] The higher the degree of mechanization, the more the contact with death is banished from life. [...] This state of indifference in killing might have become deeply embedded in the structures of our time. On a vast scale, it has only emerged during World War II, when entire human groups rendered impotent like slaughter animals hanging upside down on the production line were suppressed with an insouciance already learnt long before." (Giedion 1948, p.240)

Turning back to our cycle of conferences, "Ideas and matter", another fundamental contribution with regard to the pressing ethical issue of contemporary design is the proposal from the world of contemporary art through the *Terzo Paradiso* (*Third Paradise*) work devised by Michelangelo Pistoletto. "We live in a very precise scientific and technological age – Pistoletto states. Nowadays we are able to manipulate the DNA, and this discovery is something even more sensational than the splitting of the atom. Through the splitting of the atom, the atomic bomb was created with the resultant possibility of simultaneously destroying millions of lives. What might now be achieved through these new discoveries? *Cittadellarte* is the place where we try to combine aesthetics and ethics together. It is not only important to look at all the great innovations, which are all very interesting, since we must also understand how we might possibly use them. It is necessary to balance the manner in which this science is employed. That is why, today, we must develop at the centre of this design a sense of social ethics, a social morality, which is no longer the old "moralistic" ethics, but rather a morality that truly pays regard to the infinitely small and the infinitely big, that takes into account this huge power men wield in their hands. We are monkeys equipped with a computer. Mentally we are quite backward, we must, if not actually starting from scratch, at least start again from three, from the Third paradise. We must reconsider the larger picture, the very phenomenon of spirituality, guided and monopolized by religions that clash with each other and are an essential tool of the ideology of war. We must begin, by sifting through the entire past, and understanding what might be done for the future. This is the Third Paradise. If we do not put all of this into practice, the planet and human beings will destroy each other because of the crisis, the consumer growth and the number of persons. We must try to reach an economic and global growth balance."[7]

These topics have witnessed and continue to witness a debate between different economic and social theories, such as the so-called "economic degrowth" theorized by Serge Latouche, or the *"actor-network"* theory, a theoretical model elaborated by some French sociologists, such as Bruno Latour and Michel Callon, and by the British anthropologist John Law, or even the interpretation of contemporary society as a "liquid modernity" studied and described by the Polish sociologist Zygmunt Bauman.

Against this scenario, it seems to us interesting to have a second view at some stimulating reflections gathered by the philosopher Giancarlo Lunati, who trained at the Olivetti school of Comunità, in his book *Etica e progettualità* (1992). Lunati intends helping us exit the several ambiguities raised by the ethical issue of design "by accepting one that lies at the root of our action, consisting of a mixture of faith and reason: to believe in what we want to do, to believe in the design projects we want to offer ourselves in order to build our future day by day." (Lunati 1992, p.82) This good rule of conduct is defined by Lunati as "design ethics". "The framework within which we operate – writes Lunati – encompasses an individual discourse, of humble industriousness, even where the objectives are lofty and noble. Respect for the others must always enjoin attention not to transform suggestion into message, the rationality of the individual design into global design ideology. The difference towards the preacher must translate into a monitoring of the intentional limitation of every design. Trust in the objectives must never metamorphose into conceit: if I believe, not for that must the others necessarily believe. The desire must not be invasive and arrogant: the sense of limits must always prevail over the desire for expansion and success. Acting is not the same as unrestrained activism, and achieving one's objectives must not represent anything more than a moral commitment. A constant rule of life must however be that doing is better than contemplating, that industriousness is preferable over lethargy, that believing in ongoing improvement of oneself and society is better than resignation and cynical detestation of others and of the common well-being." (Lunati 1992, p.83)

One of the speakers who intervened at the cycle of conferences headed "Ideas and matter" was the urban anthropologist Marc Augé, known to most for his definition of "non-places", who, tackling the global topic of certain new conditions of interiority and exteriority in the "post-11 September", has highlighted the fact that "nowadays the symbolical is a form of resistance against the totalitarianism of the code. We may even think that the last opposition is the one between time and men's history."[8] Augé,

7.
See the "Third Paradise" by M. Pistoletto in this book, p.37.

in order to answer the problems of this new condition, suggests that we devote all our efforts to a necessary utopia: "education for all".

"We live in a world – Augé states – where vast gaps in wealth and culture exist. The problem is that this gap is growing daily, in the emerging countries as well. It is essential to redefine the need for a resistance. This necessitates education for all, but we know quite well that it is but a utopia towards which we may only conceive some partial progresses. We may define the school as a heterotopia of any society. Education for all is a utopia, yet a necessary utopia, without which nothing will be possible. Starting from this assumption, I would like to say something about the future: beginning with the concept that place and non-place are perfectly sympathetic and that the whole context is a planetary one, whatever one tries to accomplish is but a provisional utopia."[9]

Ending on this deep philosophical and political reflection by Marc Augé, we hope that the values of utopia and democracy merge into a renewed and far-reaching trust in the search for a reasonable relationship between ethics and design, perhaps to bring about what Edoardo Persico, quoting Dante, idealized through the concept of "substance of hoped for things"[10].

References

Flusser, V. (2003). "Un'etica nel design industriale?". In Flusser, V. (2003), *Filosofia del design*, Bruno Mondadori Editore.

Giedion, S. (1948). *Mechanization Takes Command: a contribution to anonymous history*, Oxford University Press [ed. it. (1967). *L'era della meccanizzazione*. Milano: Feltrinelli].

Lunati, G. (1992). *Etica e progettualità*. Giulio Einaudi editore.

8.
See "City-world and World-city" by M. Augé in this book, p.105.

9.
Idem, p.105.

10.
Last words of Edoardo Persico's essay *Profezia dell'architettura* published posthumously on the magazine *Casabella*, issue 102-103, June-July 1936. The original text had been prepared for the conference held the evening of 21 January 1935 in Turin, at *Società Pro Cultura Femminile* (Pro-Female Culture Society) of the Fascist Institute of Culture (Istituto Fascista di Cultura). Persico traces therein the essential lines of a history of modern architecture he intended writing. The title was supposed to be: *Oltre l'architettura* (*Beyond architecture*). It was first published in book form by Alfonso Gatto in the "Coriandoli" series of Editore Maggiani Tipografo, in 1945.

FUNDAMENTALS OF MATERIAL DESIGN CULTURE

MARINELLA FERRARA

Madec, Design Department, Politecnico di Milano, Italy

PREMISE
The international scientific community has acknowledged that the theme of the design-materials relationship is fundamental. Not only because materials are the tangible basis of objects, but also because the use of a suitable material with respect to another can make a difference in terms of a product's usability, comfort, perception, interaction, communicative value and social significance. The appropriate choice of materials is one of the key success factors for products on the market. And the designer's skill requires both technical knowledge and a broad understanding of the usability of products for end users as well as an in-depth preparation on aspects of design culture.
This theme has recently attracted a new attention in other areas – from Business Management to Human Computer Interaction – for both the added value that the design skills bring to new materials and products development, in addition to their technical value, and the influence that digital technologies exert on the material consistency of the products.
Today, a greater awareness of innovation processes is driving the re-evaluation of the role of creativity, typical of artistic and design practices - demonstrating that technical research alone is no longer able to meet the challenges for a desirable future.
What are the factors that influence consumers' product selection process? What are the elements that a designer must consider when choosing a material? According to which sequence does design intervene in the product R&D process? What is design's role in the innovation of materials? How does this role impact on design teaching methods? These are just some of the questions that have been waiting for an answer for decades. Today, after a path of gradual rapprochement to design's creative processes and user experience analysis, we are beginning to have useful answers for the advancement of design research and teaching methods.
To clarify the evolution that has led design to the current awareness of the materiality value of objects, we will refer mainly to the Italian design culture, where the areas of professional practice, research and education, while having very different rules, have a very strong relationship between them, thus influencing each other.

In Italy, for reasons mainly related to the foundation of schools within the Polytechnics and the Faculties of Architecture, design teaching has resorted to the adoption of technical skills, mostly engineering, for the study and understanding of "Materials for Design", in accordance with the name most commonly used in Italian universities. These skills oversee the study of the structure, properties, behaviour of materials and industrial processes for application. Included in Bachelor of Industrial Design programmes, engineering skills have adopted a strong materials-follows-requirements approach, which provides that the material is chosen after the definition of a design concept, so as to affect the functionality, durability, safety and the cost of the products. To support the selection process, some tools have been developed to help design students to select materials in relation to the project's technical requirements, that is, mainly with respect to the desired mechanical properties and suitable manufacturing processes to obtain the defined shapes and dimensions. CES Selector is among the most widely adopted and is based on the method developed by Michael Ashby (1999) for Mechanical Design.

After a long phase of 15 years, today design research provides a complex framework of creative design methods and material selection methods on the part of designers and of products on the part of users. It is a framework highlighting the need to update research and teaching theory, to adapt the exploration of material quality with new acquisitions, that is, with knowledge that is not derived exclusively from engineering knowledge, but from design culture, and which mainly refer to the human interaction and the social value that materials acquire in socio-economic contexts.

We shall attempt here to outline some of the key steps that have led to the development of the design culture in relation to materials. The aim is to set some concepts and definitions of a new fundamental glossary and a framework for the understanding of the current phenomena that are marking new trajectories and methods of design research in the field of materials.

1. MATERIALS FROM DESIGNERS' POINT OF VIEW

In the last decade, many authors have grappled with understanding the creative process and designers' perspective on materials. Exemplary testimonies, historical reconstructions, and analysis of design practices carried out in different operational areas of design, that is, in the field of education (Ferrara & Lucibello, 2012; de Vries, 2014) and in the professional field (Doveil, 2002; van Kesteren et al., 2007; Cardillo & Ferrara, 2008), and in the field of autonomous experimentation (Ferrara, 2011; Ferrara & Bengisu, 2013), today allow us to affirm that the design approach to the materiality of objects expresses a unique point of view and a perceptual sensitivity complementary to the technical and engineering approach. We shall attempt to highlight some of this approach's specificities.

Designers are using words like curiosity, sensory stimulation, fascination, perception, emotion, aesthetic experimentation and desire for new values and meanings to describe their personal relationship with materials. By observing the emotional involvement that shines out from their facial expressions while they touch, observe and smell the materials, it is possible to affirm that designers are given to using materials that fascinate them the most. They are interested in sensory feedback, and engage with the design theme they are broaching from the emotive experience they derive from it. Often, new materials call for a new design challenge. The creative process is launched from the physical contact and sensory experience with the new material, leading to the generation of an application idea. Therefore, materials are design inspiration factors for designers (Lefteri, 2003; Ferrara & Lucibello, 2012; Ferrara & Lecce 2015). They are the main tangible elements of expression with which designers begin to give shape to a concept.

Attracted from the outset by the sensory characteristics of the materials and by the opportunity to use them to express a personal vision of things that goes beyond the existing and the consolidated, only later do they focus on the technical aspect, considering properties, limits and opportunities arising from chemical and physical structure and production processes. In the case where the knowledge of technical limits reduces the possibilities for the application of materials, then designers take on daring design challenges to overcome or circumvent the technical and application constraints – thus, innovating the application and even the identity acquired by a material over time.

These observations on designers' creative thinking belies the affirmation that in the design process sequence, the choice of material intervenes only after the concept's definition. Instead, designers' affirmations show that the concept can be generated by the idea that springs forth from the encounter with a material. Added to this consideration is the evidence of projects that arise from the precise intent of using a new material, and the so-called projects and demonstrators often requested from designers by the materials industry with the aim of showing a new material's unexpressed potential. The creative thinking moves forward by associations, and when a new material is "discovered" by the designer, the thought immediately goes to its possible uses, in relation to knowledge of users' social needs and aspirations, and then shifts to the possible uses that take on a practical meaning in people's life.

Other valuable information on this topic stems from the direct account of certain design protagonists. The Italian designer Ettore Sottsass is the undisputed protagonist of the acquisition of the importance of the aesthetic and sensory component of materials since the 50s (Sottsass, 1954 and 1957). The inheritor of Italian and international design culture, and at the same time a critic of tradition, Sottsass deserves the credit for clearly expressing a set of new topics on sensory and mental reactions in the man-object and man-space relationship. His ideas on design as an expression of the "care for man in the consumer society" (Sottsass, 1992) led, in the 1970s, to the overcoming of the rationalist logic and iconographic baggage of the Bauhaus approach with his intellectual position based on the form-structure relationship (Sottsass, 1983). Sottsass focused attention on colours, visual patterns and three-dimensional textures, as the founding elements of the relationship that man establishes with objects and spaces, with "existential" value (Gullo, 1992), by virtue of the sensory, cognitive and emotional process that objects stimulate. Materials with their sensory characteristics assume great importance in the perception of an object and in the cognizance of spatial organisation. Therefore, materials must respond to product usability and to "perception during use", in the sense that they should encourage use and not hinder it, but to also improve it from a cognitive perspective.

While describing the project for Olivetti's "Synthesis 45" office furniture system, Sottsass said:

> "[...] the desk had to have a correct colour for reading, not too dark so that scattered sheets would not give too much of a sense of disorder and not too light so that it would not reflect too much [...].
> For filing, where big and heavy elements are typically used, we thought of a light blue colour, so as to lighten the mass.
> We gave a "wall" colour to the Cabinets – a kind of terra cotta colour –; the bookcases, which inherently are always so traditional, - the books are a kind of memory archive - we made them in brown, because it seemed to us that it would represent tradition.
> The environment took on a different colour depending on how the various functions of the different pieces of furniture were grouped. This meant that in a multi-storey building, each floor would have had a different colour, the normal obsolescence of a chromatic decision, more or less decorative, would have been less obvious."
> (Sottsass, 1983 p. 52-54).

Sottsass' approach focuses on usability through sensory effects, perception during use, user-product relationships and user-product-space because the texture, the touch, the colour or even the imperceptible smell of an object affect the interaction with it and the user's emotional state. Thus, materials that are chosen for a space may have subtle and profound social consequences. Materials that are chosen for one's home, or clothes that are worn, not only change the identity of the wearer as elements of expression, but also change the way of being and perceiving.

Although these design aspects were revealed at least forty years ago, they have never been precisely defined as design rules[1]. For Sottsass, design often involves problems that cannot be measured, but which only the designer's experience can control. We know the significance that post-modern culture has had for Italian design, giving prevalence to the hermeneutic vision of weak thought over the rational approach, according to which the design problems are not equations to resolve but necessities that reflect the uniqueness of the context and the people involved.

Ettore Sottsass' practical teaching, partly theorised by Andrea Branzi, without rules but with a deep awareness of the perceptual aspects of materials during use, generated the *Design Primario* (CDM, 1975) and the practise of CMF Design which, since the 1980s, has turned out to be like a new segment of the designer's professional activity, specialised in the "soft dimension", (i.e. the perceptual dimension rather than "hard", that is, functional and structural dimension), the "primary" or "basic" quality of the products. Facing the new perceptual dimension of materials, the traditional design tools become insufficient: a new sensitivity and a new multi-level design approach is required.

1.
Many of the perceptual aspects mentioned by Sottsass relate to the laws of visual perception, from Gestalt to the more recent findings of psychology.

CMF design is expressed in the design of colours, surfaces, material effects, namely in the elements that are recognised as the building blocks of systems and products' identity, as documented by Clino Trini Castelli in page X of this book. CMF design's development of the design field is due to this latter person, in accordance with a meta-design approach, responding to a systemic and strategic vision of the products. From 1975 to 1977, within the Montefibre Design Center (CDM), Andrea Branzi, Clino Trini Castelli and Massimo Morozzi have re-thought design methodologies and enriched them with meta-design tools. *Fisiolight, Stratitex* [Fig.1], *Fibermatching 25, Decorattivo* and *Colordinamo* manuals [Fig. 2], have been designed as tools to support professionals during the design process, choosing the soft qualities of products and spaces . The 70s were decisive for the renewal of colour theory both in relation to CDM activities, until its closure in 1978, and to Trini Castelli's activity for Industria Italiana Vernici, the research and services centre for the spreading of the Munsell system that led to the *Colorterminal IVI*, a computerized system for colour design definition by combing and cataloguing combinations of 3-colour base quantities. In addition, Trini Castelli elaborated the theory of *Qualistica* about the "perceived quality" of industrial products, which considers colour as an anti-matter, and defines a corollary of design and cognitive tools that contribute to the evolution of colour "adapting it to the design culture in the industrial world" (Moro, 2010, p. 190).

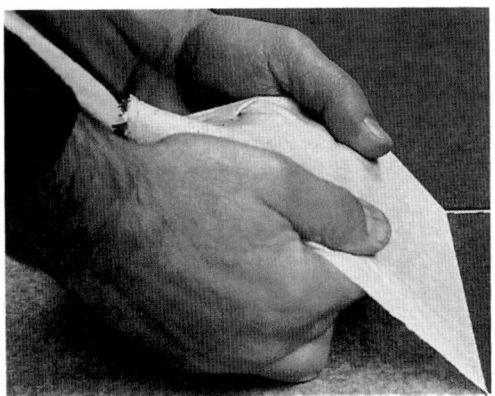
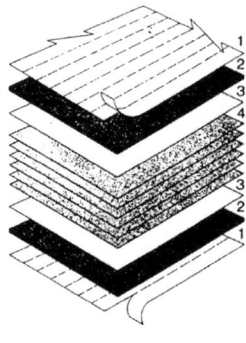

Fig.1 *Stratitex* by Clino Trini Castelli for Abet Print collection.
Photo and scheme of material layers
that compose a laminate with a retreadable textile surface, 1975.

This new design culture, experimented in Italy and then exported to the USA and Japan by the same Clino Trini Castelli, has been applied to the emotive definition of industrial products, particularly in the automotive sector for the design of colours and visual effects of the bodyworks and the design of car interiors, where materials, surfaces, textures, colours and smells are carefully designed according to the *Qualistica* method. In the 1970s, the CMF approach was applied to the design of new materials by the

Fig.2 *Colordinamo* Manual by Andrea Branzi, Clino Trini Castelli e Massimo Morozzi for CDM-Centro design Montefibre, 1976.

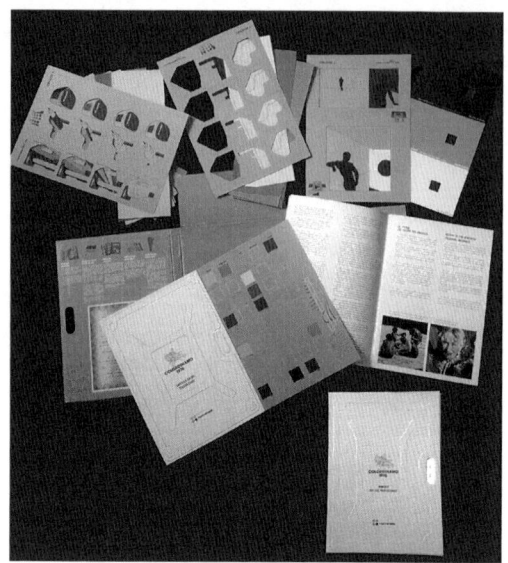

Fig.3 Abet *Lumiphos* photoluminescent laminate by Clino Trini Castelli for Abet laminate. To the left: patent (Courtesy of Castelli Design archive). To the right: an image of the layout for Eco '74 in Turin.

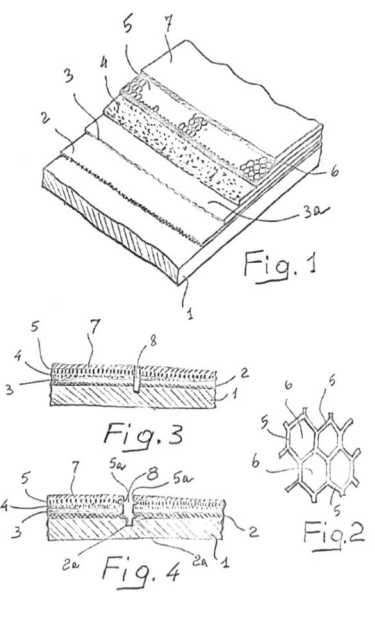

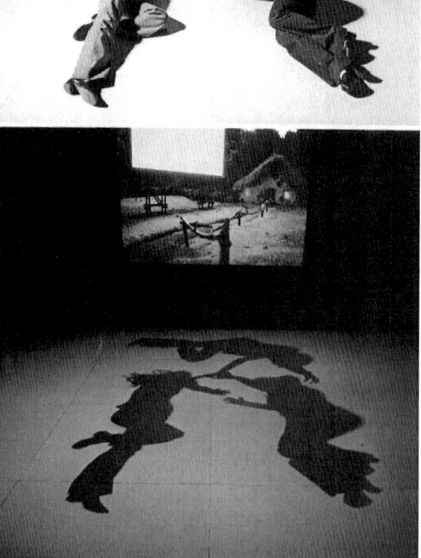

Fig.4 Abet *Relitech* laminate by Andrea Branzi, advertising, 1980.

Italian company Abet Laminati, thanks to the design collective grouped around Ettore Sottsass. Among the materials originating from design-driven research, we recall the *Print Lumiphos 14-580* photoluminescent laminate [Fig. 3], the first true case of designer-company joint development of a new material, a 1974 Abet and Trini Castelli patent which today, after 40 years, still forms part of the Abet catalogue (Lecce, 2014); the *Relitech* laminate [Fig. 4], with a particular tactile surface quality in a three-dimensional effect designed in 1980 by Andrea Branzi; the *Diafos*, the first laminate developed by the company in collaboration with Paola Navone in 1987, and awarded the ADI Gold Compass for its excellence.

Therefore, those who think that CMF design is an activity relating to the simple choice of materials for a particular application are mistaken, for it deals with perceived quality. CMF design can be defined as the design of meanings associated with the tangible aspects of products, capable of perceptively defining the intangible value of products. Through CMF design, there was a theoretical and practical shift within the discipline in the mid-1970s: the passage from the design task by "selection of materials" to "design for materials". Only in the present day has this paradigm been manifested. Sottsass' lesson remains unchanged in its validity, even if today it is preferable to replace the terms used in the 1970s with more current concepts: it is more appropriate to refer to the concept of experience "experience" rathen than to the more traditional concepts of function and perception, and similarly, by replacing the concept "interaction" with "relationship".

2. DESIGN AND THE INTANGIBLE VALUE OF MATERIALS

It is useful to turn to philosophy to address the concept of a material's intangible value. The analysis of everyday objects, in their tangible and intangible consistency, has enabled philosophers such as Baudrillard (Poster, 2002) to affirm that, even by virtue of the materials used, objects can be transformed into symbols evoking, alluding, recalling, and amplifying the specific function for which they were designed. In the perceptual process, material is revealed as a complex entity that goes beyond the functional and technical dimension, to become value, status symbol and meaning, in other words, a cultural sign capable of generating aesthetic experiences.

The same Sottsass, speaking of his work, said:

"I began to be very interested in the sensory significance of the structure's thicknesses[...] the furniture becomes a personality, it acquires its own life, independently of all the rest of the surrounding environment[...].
Rather than neutralising the furniture, I loaded it with meanings[...] any form is always metaphorical[...]" (Sottsass, 1983 p. 72).

Colours, smells, weight, tactile effects, response to light and sound, are all characteristics of materials that open up to a horizon of sense in human receptivity that is both active and passive: *Trans-possibility* and *trans-passibility* as the philosopher Henry Maldiney (1991) defines it. Penetrated into the different layers of society, and once accepted by large groups of people, these characteristics are automatically transformed into cultural signs producing aesthetic experiences. The aesthetic experience that is derived forms the basis of the relationship between user and product which underlies the choice of consumer purchasing.

Besides those from philosophical and semiological studies, interesting interpretations on this topic come from communication design area of study (Krippendorff, 2006) and design management (Verganti, 2008) which have defined the design-driven innovation approach, based on the study of the innovation of meaning proposed by design-oriented Italian companies with their products.

Added to these studies is the historical research that focused on design functionality in collaboration with materials manufacturing companies, to highlight designers' ability to generate the innovation of meaning of industrial products, through the treatment of tactile features. Italian design history is rich with stories of collaboration between designers and companies established with the specific intent to dedicate themselves to a new technological challenge, such as that of plastic materials. It ranges from the company Pirelli, founded in the second half of the nineteenth century for the manufacture of rubber products to more recently established companies such as Arflex, founded in 1950 by a Pirelli manager with the goal of using foam rubber for the production of upholstery; B&B of 1956 and Gufram, that exploited the development of polyurethane foam to industrialise upholstered furniture; Kartell founded in 1949 by Giulio Castelli, a chemical engineer who aspired to create something good with the plastic materials developed by Natta's studies, and whose applications were still unexplored; and Abet Laminati to which we owe the development of plastic laminates (Lecce, 2014); and so many more. The introduction of synthetic plastic materials between the 1950s and 1960s technically and functionally improved industrial production with increased performance of lightness, unbreakability, resistance and durability. Starting from new products, from tableware to furnishings, they created new markets and distribution channels, and a new horizon of sense opened up for consumers, new aesthetic experiences with pop items, ironic, vivid colours and at an attractive price.

Some research deserves credit for re-enacting design micro-stories by contextualising, describing and documenting the design process of some of the many icons of Italian design. And here, we are referring in particular to a study promoted by Madec in collaboration with AIS/design, the Italian Association of design historians, which has recognised and defined an *Italian Way to Material Design* based on the awareness of the perceptual aspects of materials (Bosoni & Ferrara, 2014) and on the ability to enhance the communicative value of products through a process that has been defined as the "leavening of material" capable of eliciting synaesthesias with the treatment of tactile surfaces and appropriate associations of sense, which generate emotions in consumers (Ferrara, 2014). One case for all, marked as original in the history of Italian industrial design, is that of Pirelli products for hygiene and the home. Between the 1940s and the 1970s, these rubber products with a rubber surface design won over consumers in Italy and in various European countries. Tactile and visual effects, particularly in hot water bottles [Fig. 5] and children's hygiene products, - soft texture from warm colours, alluding to velvety skin - amplified the sensation of delicacy and intimacy, predisposing bodily contact with these items. Overall, the tangibility aspects (forms, materials, etc.) plus visual communication artifacts, enriched with metaphorical meanings, contributed to the creation of empathy, by transmitting sensations of softness, warmth and comfort. Loaded with narrative value, they succeeded in arousing positive emotions in the consumer/user (Ferrara, 2014).

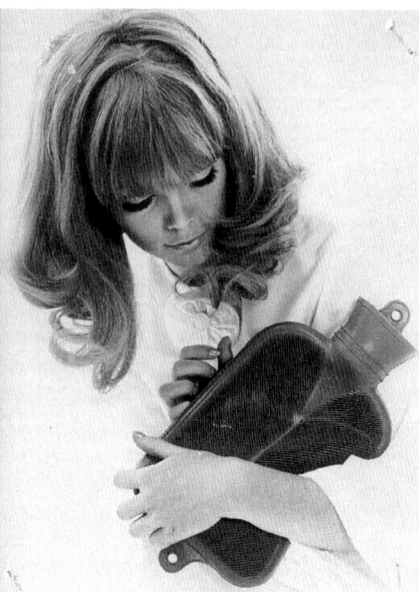

Fig. 5 Pirelli hot water bottles from 50s to 70s. From left to right: Photo for advertising, 1949; Photo by Papafava, 1968; Photo of new models, 1954; Poster by Lora Lamm, 1959.

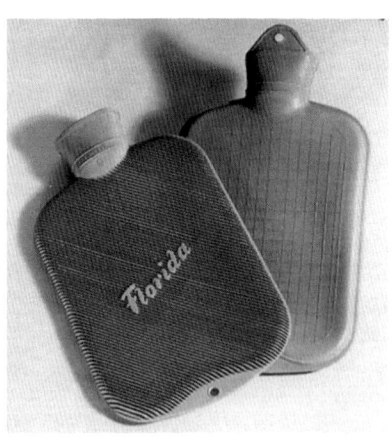

In creative dynamics, as in perceptual ones, the expressive language of shapes and materials, in association with the verbal language of communicative artifacts, enables the creation of sense connections that become dominant elements in the construction of a shared desirable fantasy, even in cases of a new interpretative model of reality. This thesis is reflected in Jean Baudrillard semiological theory who, in his 1968 essay *The system of objects* states: "[…] in commodity the relation of word, image or meaning is broken and restructured so that its force is directed, not to the referent of use value or utility, but to desire." (Poster, 2002)

In Pirelli's specific case, the creative process has been influenced by the exchange of ideas between technical and humanistic culture stimulated by the house organ "Pirelli. Magazine of information and technique", where renowned intellectuals, engineers and designers published the technical performance of new products. The narration enriched the experience of products with metonymic and metaphorical meanings which alluded to recondite meanings (Ferrara, 2014). Thanks to designers' propensity to interpret materials in ever new and unexpected ways, by suitably dosing functionality, sensoriality and significance, each new material acquired a particular identity, capable of being renewed at each new application in everyday objects that inverted consolidated perception.

3. MATERIAL DESIGN STRATEGY

A fundamental experience in the way of defining the Italian Material design culture (Bosoni & Ferrara, 2014) was carried out between the mid 80's and the 90's at Domus Academy, a well-known Italian school of design of advanced level (Master), with international profile, founded in Milan in 1983 by Maria Grazia Mazzocchi.

Since its foundation, DA has been equipped with a research center, directed by Ezio Manzini from 1983 and by Antonio Petrillo from 1990 to the late 90's, which has brought in a fluid mix of teaching and research activities about "new materials", by the most qualified professionals and design scholars. Among them Andrea Branzi, Clino Trini Castelli, Anna Castelli Ferrieri, Alberto Meda, Gaetano Pesce, Denis Santachiara and many more. Since the mid-1980s, thanks to a close connection between the research center, strategy consultants and enterprises management in the field of materials, the DARC was a laboratory where to explore the possible productive and expressive applications of new materials and upcoming advanced technologies that in the industrial context. We must remember that in the 80's, plastic production had surpassed that of steel, supplanting many other traditional materials, thanks to the development of high-tech materials (HPP) that provided high performance. Almost at the same time, the oil and gas crisis in the 1970s and 90s have been forced many companies to review their production to start a lengthy experimentation phase of new plastics recycling technologies and the production of new materials from renewable sources.

In the 1990s, with the leadership of Antonio Petrillo, the orientation of the DARC was influenced by: "taking notice of some profound transformations of industrial society, particularly the progressive transition from the commodity society to the services society, the marketing crisis as a market reading system and as the only form of orientation for business initiative, and the growing importance of corporate design. Antonio Petrillo used to say that a move from a system of circulation of goods and consumer goods to a global communications circuit was happening: the information and cultural content contained in the product became an integral part of the product enjoying by the people." (Ceppi, 2014) As a result, designers had to on two levels - the company and the products. The DARC was thought to be a support structure for professionals to contribute to building business strategies, orienting enterprise production philosophy and product proposals.

Today, the DARC retrospective analysis allows us to trace a design method that continues with the experiences of the Design Primario (1975), capable of defining the strategic aspects of material research, according to a broad and coordinated concept of Design operation between product design and visual design communication skills that generate material and immaterial quality. "The DA approach represents a unique approach in the 90's, intended to be referring to future generations of designers." (Ceppi, 2014)

Among the projects developed by DARC, we would like to remember *Neolite*, the research for the development and launch of the first heterogeneous recycled plastic produced in Italy. The project was started in 1990 in cooperation with Montedison, RPE/Montedipe, Assoplast, and CSI Research, which managed to handle two different plastic families so that they could be combined in a single dough, then liquefied and reduced to granules. In DARC this research, like many others, has been developing at the level of business consultancy to first define the philosophy (the design vision) of the material, thus sensitizing the company's internal frameworks on the new themes of culture and environmental sensitivity that were developing at a social level, and to program research initiatives for material applications and visual communication-related.

The recycled plastic material that emerged as waste led to the strategic choice of defining a new identity of the material, in line with the emblematic features of a new environmental culture, and a set of potential applications that would convey and promote the material qualities (technical properties and sensory and aesthetic characteristics).

The research consisted in designing the material concept, what we would now call 'material vision', and define the semantic and performance qualities of the material, develop technical research, applications, and a communicative strategy.

From the technical point of view, the exact formulation of plastic components was defined for some material variants and processes. From the point view of the sensory characteristics, the optical and tactile qualities were designed, and the couplings with other materials were introduced through the insertion of charges, such as metal powders, which ennobled the forms of aging over time.

The material vision was related to an adequate communicative strategy, which began with the design of the material name. *Neolite* is, in fact, the term coined by Andrea Branzi and Pierre Restany to convey the concept of the beginning of a new era of materials. Keywords were chosen to transfer the philosophy of recycling, not just as a form of technical engineering, but also as a manifestation of a changed cultural attitude. Consequently, they have worked with the methods of product design to develop new products related to the philosophy of material, on which to experiment with the technical and aesthetic potential and to define a "brief" for new collections [Fig. 6]. This demonstrates that a fundamental part of Material Design's research goes through a product-focused innovation demand, where seeking new aesthetics in accordance with the material philosophy ensures innovation in solutions.

The methodology outlined in this and other DA research included 4 types of highly integrated and related activities: Concept Design, Strategic Positioning, Product Design and Visual Communication.

The concept design phase is the one in which an integrated image, communication, and product development strategy are conceived to allow the material to be positioned at a quality level appropriate to the defined market segment.

The strategic aspects of communication and strategic positioning are intended to help end consumers to better understanding of their quality and performance. The communication was not only aimed at the B2C consumer market but instead turned to the world of producers and transformer companies, starting from the need for raw material formulators and process transformers to understand the actual technical or aesthetic potential of their potential clients.

The research became the premise for the realization of an exhibition that expressed the new cultural orientation in the production and consumption of plastics. "Neolite-Metamorphoses of plastics" at the Triennale of Milan, was one of the first exhibition of materials organized in a place other than the fair exhibition spaces traditionally used by the industry. The exhibition presented to the public new materials, various communication artifacts, projects and prototypes of potential applications, including the projects developed for the Replastic Consortium on the subject of separate collection of waste in outdoor applications.

In the exhibition, the theme of material design was directly linked to the theme of sustainability, new productive processes, new products and new behavioral logics such as differentiated waste collection. The different levels of the project intertwined with

Fig.6 Neolite texture tests, 1990.

the desire to stimulate and grow in consumers a more mature sensitivity to materials. The exhibition, with Bruno Munari's layout, defined the elements of a new ecological imagery with corresponding aesthetic qualities, through the development of a global scenario in which the different forms of plastics recovery were ordered and illustrated the future possibilities for application development. The exhibition followed in a new style of advertising communication of materials.

References

Ashby, M. (1999). *Materials Selection in Mechanical Design* (3rd ed.). Burlington, Massachusetts: Butterworth-Heinemann.

Bosoni, G. & Ferrara, M. (eds)(2014). Material Design. Learning from the History. *AIS/Design. Storia e Ricerche*, 4:1.

Cardillo, M. & Ferrara, M. (2008). *Materiali intelligenti, sensibili, interattivi*. Milan: Lupetti editori di comunicazione.

CDM-Centro Design Montefibre (ed) (1975). Il Design Primario. *Casabella*, 408: 41-48.

Ceppi, G. (2014). Il design dei materiali in italia. il contributo del centro ricerche domus academy 1990-1998. AIS/Design. Storia e ricerche, 4:0408.

Doveil, F. (2002). *iMade: l'innovazione materiale nell'industria italiana dell'arredamento*. Milan: Federlegno Arredo.

Ferrara, M. (2011). Design and self-production. The advanced dimension of handcraft. *Strategic Design Research Journal* 4 (1), 5:13.

Ferrara, M. (2014). "Lievitare" la materia. Pirelli, la gomma, il design e la dimensione politecnica. *AIS/Design Storia e Ricerche*, 4(1), ID:0401.

Ferrara, M. & Bengisu, M. (2013). *Materials that Change Color. Smart Materials, Intelligent Design*, Berlin-Milan: Springer - Politecnico di Milano.

Ferrara, M. & Lecce, C. (2015). MADEC. Material Design Culture. In E. Duarte, C. Duarte, F. Carvalho Rodrigues (eds), Senses & Sensibility '15: Design as a Trade, proceedings of the UNIDCOM/IADE 8th International Conference (pp. 490-497). Lisbon: IADE-Creative University.

Ferrara, M. & Lucibello, S. (2012). Teaching material design. Research on teaching methodology about materials in industrial design. *Strategic Design Research Journal* 5:2, 75-83.

Kesteren, I.E.H. van, Stappers, P.J., Bruijn, J.C.M. de (2007). Materials in Products-Selection: Tools for Including User-Interaction Aspects in Materials Selection. *International Journal of Design*, 1-1: 41-55.

Krippendorff, K. (2006). *The Semantic Turn. A New Foundation for Design* London. New York: Taylor & Francis.

Lecce, (2014). Abet Laminati: il design delle superfici. In Bosoni, G. & Ferrara, M. (eds)(2014). Material Design. Learning from the History. *AIS/Design. Storia e Ricerche*, 4.

Lefteri, (2003). *Ceramics: Materials for ispirational design.* Mies - Switzerland: RotoVision.

Maldiney H. (1991). *Penser l'homme et la follie.* Paris: Millon.

Manzini, E., & Petrillo, A. (eds) (1991). *Neolite. Metamorfosi delle plastiche.* Milan: Domus Academy.

Moro, M. (ed) (2010). Clino Trini Castelli e il Design Primario. In C. Boeri (ed), Colore. Quaderni di cultura e progetto del colore (pp 190-207). Milan: IDC Colour Centre.

Poster, M. (2002). *Jean Baudrillard: Selected Writings.* Palo Alto: Stanford University Press, p1.

Sottsass, E. (1954). Struttura e colore. *Domus*, 299: 47-48.

Sottsass, E. (1956). Per un Bauhaus immaginista contro un Bauhaus immaginario. *Casa e Turismo*, 12: 15-18.

Sottsass, E. (1957). Struttura e colore. *Domus*, 327: 20-22.

Sottsass, E. (1983) Disegno di mobili standard. In A. Martorana (ed), Storie e progetti di un designer italiano. Quattro lezioni di Ettore Sottsass Jr. (p 52). Firenze: Alinea.

Gullo, L. (1992). (2013, Oct 8). TGR RAI - Lillo Gullo intervista Ettore Sottsass [Video File]. Retrieved from https://www.youtube.com/watch?v=Cmu0m5NorF4.

Verganti, (2008) Design, meaning and radical innovation: A meta-model and a research agenda. *Journal of Product Innovation Management*, 25(5): 436-456.

Vries, M. J. de (2014). The Context Approach to Learning Material Properties in Design (-Related) Education. In Karana, E., Pedgley, O., Rognoli, V., *Materials Experience* (pp 329-336). Butterworth-Heinemann: Elsevier.

SHIFTING TO DESIGN-DRIVEN MATERIAL INNOVATION

MARINELLA FERRARA
MADEC, Design Department, Politecnico di Milano

In the last 15 years, materials research has assumed new strategic importance in the intersection between science and society (Ferrara, 2009; 2010; 2015), so much so that it is one of the most funded and flourishing scientific research fields. The coming years will be dedicated, even more intensively than past ones, to the tactile reconversion of materials products. On the one hand, progress in research is envisaged for enabling, smart, advanced and augmented solutions in strategic sectors such as life sciences, and to increase the added value of products and services with new features, such as interaction and connectivity. On the other hand, there has been an advance in the search for bio-based solutions that are increasingly compatible with nature, capable of replacing petroleum-derived synthetic materials, to return at end-of-life in the natural cycle and meet the global demand for sustainable industrial products.
In the face of the progressive advancement of scientific and technological research, design is a strategic competence if placed at the service of desired changes. Thanks to designers' ability to design the characteristics of the materials, by dispensing sensoriality, and significance, it is possible to create deep emotional connections between users and products so as to guide consumers towards more sustainable materials. Not only during the design of a product, but also with the aim of conceiving new material concepts, according to a complementary approach to materials scientists.

1. DESIGN-DRIVEN MATERIALS AND MATERIAL THINKING
Since the beginning of the century, a strong environmental sensitivity has stimulated designers towards research aimed at the use of waste materials applied as raw materials for new productions. A case in point is the German designer Tobias Juretzek who, by using old garments (trousers, T-shirts and denim) as material for new products, he soaked them in a mixture of resins to then compress them at high pressure. This production process interested the Italian company Casamania which produced the *rememberMe* collection of furniture with this technique [Fig.1].
Another example is Massimiliano Adami's research who completely autonomously experimented with a production process to create furniture from scrap plastic materials. The technique consists in incorporating waste plastics (bottles, toys, and scrap televisions and computers) in a polyurethane foam casting. After having assembled the pieces in a formwork and made the casting, once solidified, the designer skilfully

practiced some cuts on the surface of the pieces, and highlighted the shapes and the recesses of the plastic waste, revealing coloured and differently textured sections of the discarded objects, which thus acquired a strong expressive value [Fig. 2]. The significance of the tested techniques from these designers lies in the ingenious transformation of waste into a valuable resource in art & design practices.

Fig. 1 - Tobias Juretzek, *rememberMe* collection, for Casamania.

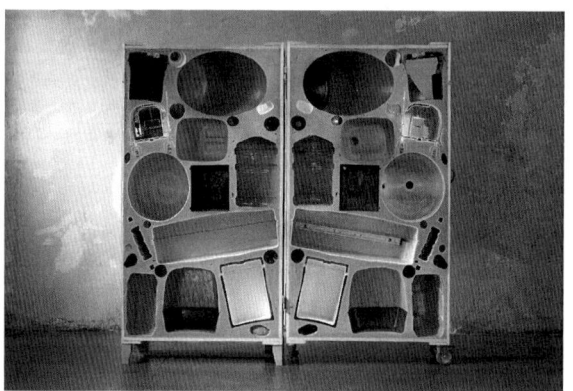

Fig. 2 - Massimiliano Adami, *Fossili moderni*, self-production from 2005, and *Gommapiuma Decor 2007*.

Many other creative practices have been experimented to produce materials derived from agriculture or food waste, applying conventional technologies and analyzing the organic matter behaviors at the macro or micro scale.

Another symbolic case is Maurizio Montalti's research, exhibited by the same author on page 71 of this book. This research, which moves between design and molecular biology, is symbolic of an interesting phenomenon, the so-called *Material Activism*

(Ribul, 2013) which identifies an active movement of creators and makers that explore new "material visions" with low-tech methods and propose new models of cross-disciplinary research and collaborative practices aimed at the design of systems for materials production. By material vision we mean a way of conceiving, of interpreting the material which involves a choice of perception in the experience, corresponding to the meaning that the material assumes during use for the end user. Thus, Maurizio Montalti pursues the vision of "growing material", emphasising the concept of a renewable material which becomes a self-production tool, to be modelled, stabilised and used, to then be reintroduced into the production cycle of Nature. The vision also coincides with a precise narration strategy, which in this case serves as an expedient to bring the user closer to a living matter, the mycelium, commonly associated with the negative concept of mould and spoilt food, to perceptually replace it with the meaning of "cultivation" and growing new materials in precise productions such as that of blue cheeses and of antibiotics against bacteria. As already discussed, the phenomenon of research on new self-produced materials, developed in relation to the wider phenomenon of the rediscovery of the American *Do It Yourself* culture, expresses the desire of young designers to recapture technological activities as 'facts of material culture' (Maldonado, 1976 p.16), spreading new meanings on technique, matter, its use and the circular economy (Ferrara, 2014). Beyond the ideology on the democratisation of production, on which this phenomenon rests in some cases, the pragmatic intent of *Material Activists* is to spread a new relationship with technological processes and materials, potentially useful for *empowering society* [Fig. 3]. Design opens to the collaboration with life sciences, and hybridises with skills from other fields of science such as biomimetics, digital computing, chemistry, molecular gastronomy, etc. Many project are thought-provoking design approaches providing key insights into how material will be utilised to shape our future environments, as declared by Jenny Lee (2015) in her book *Material Alchemy: Redefining Materiality Within The 21st Century*. There emerges a new designer-scientist figure: alone inventor who creates with his own hands, takes advantage of *technological expertise* and the network knowledge, and communicate his experiments with a personal narrative style [Fig. 4]. Exploring key topics such as synthetic biology, or how new technologies such as

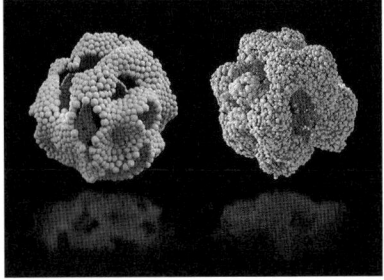

Fig. 3 A study into a porous, lightweight and permeable substrate of concrete mixture based on cell division patterns: Project by Taehyun Terry Lee, Wen Cheng, Dan Lin, Sul Ah Lee, Supervisor Prof. Marco Cruz, Richard Beckett, Javier Ruiz, The Bartlett UCL, 2015. From Materiability website.

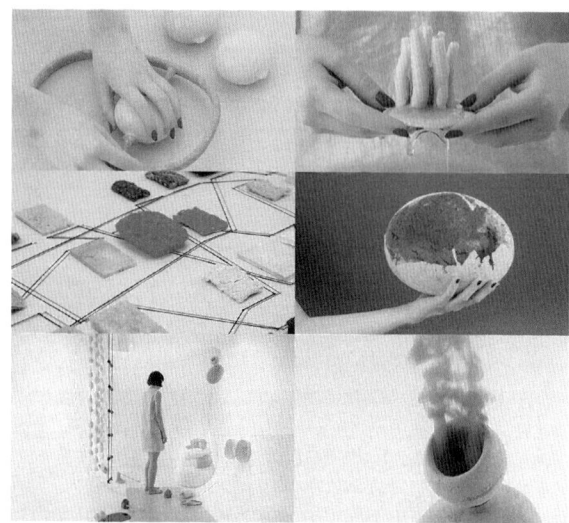

Fig. 4 Johanna Schmeer, Bioplastic Fantastic project. She investigates new types of products and interactions that might emerge from material innovations in the fields of bio- and nanotechnology.

3D printing are revolutionising the manufacturing industry, designers and maker with different background are proposing new insights to explore materiality to showcase new responses to material innovation and providing key insights into how materials will be utilised to shape our future environments.

Another interesting dimension to be included in Material Activism, albeit with different aims compared to the previous examples, concerns the autonomous experiments developed within the intersection between design and human-computer interaction (HCI), such as those of Marcelo Coelho [Fig. 5] and Andrea Minuto for the development of tangible interfaces [Fig. 6]. These researches contribute to "invent" new materials through the integration of conventional materials with microelectronic devices and smart materials distributed in the material, such as the so-called *actuated matter*, as per the term used by Manuel Kretzer in his Materiability network[1], or the *augmented materials* "systematically programmed to augment the behaviour of synthetic materials" (Razzeque et al., 2013). The value of speculations on possible new "composite matter-related systems" with advanced functionalities, understood as systems resulting from the aggregations of materials and devices potentially serving to confer tangibility on man-object interfaces, is a pregnant one: surfaces become sensitive and intelligent, acquire an active and reactive behaviour, with kinetic, acoustic and visual response capabilities (deliberate traces), to connect you to computer networks, transfer and receive information. These new performances transform the materials into communication and interaction tools, more than was the case in the past. Moreover, they open up new possibilities for perceptive and emotional involvement, since

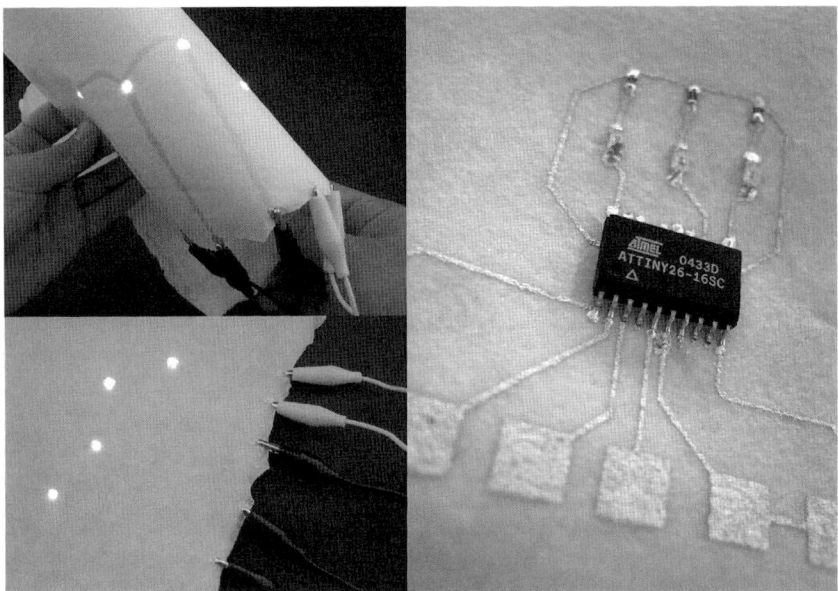

Fig. 5 Pulp based Computing and Fabrique based Computing by Marcelo Coelho, 2009.

they incorporate aspects of the organic systems to such an extent as to be possibly defined as "live objects".

All the examples previously described of applied design to new production processes and related materials, although having many differences between them in terms of development methods and the productive result obtained, are part of a phenomenon that is today very widespread in Europe, operating a translation of materials research from methods and expressive languages of the science of materials and engineering, to methods and expressive languages of design. Research is shifting its focus onto the material from the performance of materials to their meaning. In fact, the performance of materials in the three cases presented here, do not show increased performance. Research places emphasis on the significance that the material assumes in the social context and on the emotion that the material provokes on users sensitive to the theme of environmental sustainability. Conversely, emphasis on the formal research of applications decreases, taking archetypal forms as a reference, in a way that recalls the No-form approach defined by Clino Trini Castelli (p. 45).

The observation of emerging and widespread creative practices focused on materials, confirms the enlargement of the areas of design competence towards the design of materials, processes and production systems. The resulting materials are, in our

1.
http://materiability.com/

Fig. 6 Experimental model of an "Emotive Environment" by Actuated Matter Workshop, IAD Interaction Design, Zurich University of the Arts, 2011. The workshop explored the application of smart materials in architecture with respect to their ability to transform architecture into a space that relates to its inhabitants in an emotive and responsive way. From Materiability website.

opinion, definable design-driven materials, that is, new materials that are derived from design research, characterised by the purposes, methods and expressive languages typical of this discipline. Design research configures the characteristics of the materials, the production methods, and at best, application scenarios and above all, the meanings for which materials are the bearers through precise material visions. Design has a fundamental role in the design-driven material innovation process: to understand the socio-cultural values and to interpret them in innovative materials and products whose sensory characteristics are capable of generating significant user experiences, that is, in line with cultural and lifestyle changes, that affect the aesthetic as well. By taking advantage of these practices and the availability of new knowledge and technologies, design research tends towards radical product innovation, through the innovation of the meaning for which the material is the bearer. We can talk about a new creative strategy coming from art practice, that can be termed "Material Thinking" because it refers to the Design Thinking approach to be applied during the design process of a new material and production processes.

At present, practices of design-driven materials, often carried out under autonomous experimentation and in the absence of any connection with the industry, find several difficulties for effective industrial application. This is especially true when the research is limited to the conception of a material, and omitting the design of applications. Yet, it is not excluded that a change of the technical-productive paradigm in the near future can integrate the Material Thinking strategy and thus design-driven material approach with R&D industrial processes.

With this objective, and on the occasion of the *Design for Enterprises*[2] European project, financed by the EASME (the Executive Agency for Small and Medium-sized Enterprises), which aims to increase the innovative capacity of small and medium-sized companies in Europe, we conducted a detailed study leading to the *Design for Materials* course. This course "clarifies the several factors to be considered for developing or choosing a material when designing consumer products" (2016). Design for Materials looks at materials from a different perspective when compared to past engineering thinking: material performances are also based on sensory perception, on consumer experience, and on cultural values. Composed of five sections (Why design for material?; About material performances and user emotions; Design contribution to materials research; Creativity-driven innovation material methodology; and a selection of materials that will change the future), the course helps enterprises to understand that material innovation requires both technological exploration and a broader understanding of its meaningful application for consumers. It explains that material performance is also based on sensory perception, consumer experience, and cultural values. That it becomes strategic to foresee innovation trends impelled by social, cultural, economic and environmental drivers. In every section, the course includes case studies of small and medium-sized enterprises that are applying design in their R&D process, and are achieving both product functionality, material distinctiveness and clarity of message. The course aims to teach how to manage a design process where different actors, such as scientists, suppliers, creative communities and consumers are becoming deeply engaged.

"Design for Materials", understood as the advancement of CMF design, presents itself as a strategic competence that increases the innovative capacity of businesses useful for guiding the product innovation trend towards the development of higher-performance materials, with less impact on the eco-system and as bearers of new social meanings. The course also acts as a guide for new design-driven businesses based on the development of new materials. In fact, it presents some case histories of start-ups that have applied the creative approach by developing new materials. This is the case of "Woodskin", a young Italian SME that patented a new composite material based on wood, combining the rigidity of traditional material with the flexibility of textiles, and enabling different applications such as customisable products and architectural elements. Since its composite material's first patent, the start-up has been developing new material combinations for specific and different applications.

2. DESIGN-DRIVEN MATERIAL INNOVATION

Today, we are witnessing an important step forward in the recognition of design competencies in the field of innovation. According to the European industrial policy on research and technological development, the integration of design competences becomes a priority because it contributes to "closing the cycle of innovation" (2010, the European Competitiveness Report). The generation of ideas for new products and

2.
www.designforenterprises.eu

services according to social and technological trends is one of the main challenges for the SMEs, above all in the clusters of furniture, building and mechanics.

There are several innovative materials which, once developed by scientific laboratories, encounter difficulties in their application to new serial productions. The materials libraries, established in connection with the hyperproduction of new materials, have attempted to reduce the gap between producers and applicators. But this is only a small wedge in a very wide gap. If, instead, technical applications of excellence were to be developed together with new materials, the chances of success for the new materials would increase, because the time of industrial adoption and the time-to-market would be considerably reduced.

Starting from these considerations, design research has been indicated as a solution for innovation's critical node, that is, the length of time it takes. In fact, given the time needed for a new material's technological and industrial development (which includes experimentation and the material's characterisation, the development of the process and the production chain, the creation of a network of suppliers, the achievement of a good level of production efficiency and obtaining the necessary certifications), and the market entry of the new material that must find suitable and specific applications, the innovation process takes twenty years on average.

To reduce innovation times and increase the new materials' chances of success in order to accelerate the conversion of production towards materials that are more sustainable and efficient than those that we currently use, it is considered useful to have a strategic focus of design ahead of the R&D phase, and placed in the early stages of a new material's technical development. Similar planning would overturn the method adopted to-date, which assigned designers the role of choosing the material in the post-technical development phase. This method, as we have already mentioned at the beginning, was defined at least twenty years ago by engineering skills that only considered the required design practices in the final stage of a new product's technical development, as an opportunity for the aesthetics of products through the product's formal definition irrespective of the technological content.

Instead, by being placed in the early stages of R&D, design research contributes to the choice of "what to design" on the basis of "why design it", by directing innovation towards the understanding of "subtle and unspoken dynamics in socio-cultural models and results in proposing radically new meanings and languages that often implies a change in the socio-cultural regimes" (Verganti & Dell'Era, 2014, p. 146). Indeed, we should remember that innovation is not only a techno-scientific fact, but a social process of dynamic nature that is often accompanied by other forms of transformation that may relate to the aesthetic and symbolic characteristics of the materials, their ability to fulfil latent social needs, to emotionally engage the consumer and to generate new product meanings. In the R&D phase, design discourse with Material Thinking strategy shifts the focus from the technology-centred approach to the human-centred approach that takes into account the user-material/product interaction. The designer's unique know-how, in fact, lies in the design of a "meaningful" interface of products: easy to understand and use, and pleasant in the experience. Meaningful experiences are those that develop markets, by also introducing radical innovations that originate from interpretations of the profound changes which we are experiencing.

When a new material emerges, it incorporates many potential applications: some are promoted by those who initially guided the technological development, aimed at replacing old materials to improve the performance of existing products; but there are applications that require imagining new forms of consumption, solutions for new needs and experiences of use. With design's creative contribution, new materials can provide suitable answers to new consumer aspirations and can accelerate the entry of innovative materials into the market.

For this reason, today the priority is to foster an effective penetration of design research approach in companies' R&D processes, with the awareness that the relationship between research and innovation will be increasingly correlated and dependent on cross-disciplinarity.

The current frontier in product innovation is based on KET (Key Enabling Technology) that include nanotechnologies and applications for advanced, smart, as well as eco and bio-materials. These are often far from obvious in the context of microscopes and lab benches and, at the same time, are quite distant from the resources of every freelance designer, usually working on a DIY basis and with a traditional, instead of a strong technical, background. For this reasons, one of the biggest challenges about innovation for the future of European industry is to fill the gap between fundamental material science and its applications, anticipating design contribution in the material innovation process, while gearing towards developing product ideas based on scientific work with materials. These setting processes will be based on strong interaction and hybridization between science (rational-deductive-analytical) and design (empirical-inductive-synthetic) methods, in order to generate disruptive innovation, based on materials both in terms of performances and aesthetics and meaning.

3. OPEN MATERIAL INNOVATION AND DESIGN ROLES

Over the last decade, several issues have limited innovative processes: the onset of the economic crisis, the intensification of global competition, the reduction in the life cycle of new products, and the difficulty of sustaining increased R&D spending. In this challenging context, innovation networks and Open Innovation theory, with their many points in common and both based on the role of knowledge, have become the leitmotif of management studies, in addition to the goal of strengthening current businesses and building new ones.

Innovative Networking Theory (Powell et al., 1996) is based on enterprise resource vision, internal distinctive skills, dynamic and relational capabilities, and on external process management.

Open Innovation is the organizational paradigm for industrial enterprises defined by H. W. Chesbrough (2006) as a model for accelerating innovation processes. According to the US economist, Open Innovation (OI) consists of: "... the use of purposive inflows and outflows of knowledge to accelerate internal innovation, and expand the markets for external use of innovation respectively". Unlike the traditional vertical integration model of knowledge in R&D processes, OI indicates the need to create an open ecosystem for different knowledge flows and exchange of ideas. In order to increase the company's innovative capacity, it is necessary to enrich the internal knowledge base within the company through external knowledge: from research centres, suppliers, business partners and spin-offs, even competing companies, customers and end users, markets, sales networks, and companies. The concept on which the model is based is that the added value that different companies can give by working collaboratively and proactively is greater than the sum of individual companies' added values. This leads the company to be more permeable and to expand its interests beyond the channels of its own business, allowing innovation processes to move and develop without barriers. In the spirit of Open Innovation, many creative industry players have begun to demonstrate that design has a role in accelerating innovation and in the appropriate orientation of innovation. "Within the firm, design is revealed as a core capability that shapes open innovation practice, reflecting its role in innovation partitioning." (Acha, 2008 p. 5). In a number of areas, designing operations are appreciated not only as inputs in the innovation process but also as skills that facilitate the conduct of open innovation activities (Acha, 2008). Acha's statements are based not only on the analysis of literature on management and innovation issues, but also on a study on the innovative performance of UK companies dealing with highly complex products and systems (high added value, customization and intensive use of software) and operating in a variety of industries (from construction to industrial machinery for food, textiles, etc.). According to Acha, the more complex the task the more design plays an instrumental and strategic role, because it is necessary to facilitate the interaction of the various parties involved in the open innovation process by making sure these, even if they operate independently, are coordinated with respect to a common goal. The role of design is fundamental in translating the expectations of the organizations concerned into an understanding of the various parts involved.

On other occasions, designers have been defined as the "facilitators of innovation processes" (Krippendorff, 2006). Design has demonstrated the ability to mediate between different knowledge, and on how many different operators contribute to the process of product creation (from the industrial entrepreneur to the artisan supplier). And they have demonstrated the ability to be interpreters of consumer needs, by mediating the relationship between the production system and the consumer system. Design research has developed participative, collaborative, and open design approaches that manage processes where multiple companies, clusters, and know-hows can be involved, according to a multilevel design model.

In collaborative design (*co-design*), participants have different backgrounds and views and represent different interests. The design process results from the interaction of a variety of disciplines and stakeholders, including design experts and finals users who contribute their ideas and take action. Due to differences, mutual understanding loses effectiveness and slows down the process. Instead, the process should flow to foster cooperation between the various actors (suppliers, manufacturers, industries and traditional SMEs) and the cross-fertilisation of knowledge to generate new knowledge. Here, we expose how *design culture*, with its skills and capabilities, is able to facilitate highly complex socio-technical processes, such as design-driven innovative processes applied to the material production sector:

- The designer tends to facilitate a socio-technical collaboration process through seeing, listening, understanding and interpreting the exchange of information between the design process' participants and collaborators, while placing the

user at the centre of the design process; they propose their visions on the basis of their interpretation of problems to be overcome and opportunities to be gained;
- Dialogic capabilities combined with the *material thinking* approach, support the process of changing an existing situation into a new desirable situation. As regards a problem-solving task, design capabilities are able to involve the right team work between different actors, combining the often contradictory desires and interests in an integrated design of a new socio-technical system, thanks to the ability to develop a common thinking framework, i.e. a shared material vision. They can identify interviewees who are suitable for the production challenges, and involve them in collaborative projects. Designers can specifically design a multidisciplinary project team to promote knowledge transfer and process management to ensure independence of the various tasks;
- As experts of the production chains and the various production sectors, designers manage to operate across production sector boundaries and skills in order to foster interaction. They are skilled in transferring technologies from one sector to another and they understand the processes involving raw material suppliers, commodity converters, product users, and so on;
- During the innovative process, visual communication capabilities allow the communicative process of participants to foster mutual understanding so as to facilitate trading by enhancing differences and by communicating points of view in a comprehensible way for all participants. *Envisioning* activities makes it possible to overcome the diversity of specialized languages in the interaction by using visual language to build a shared vision and a common vocabulary. As a visionary mediator, the designer can reconcile the interests of the parties, including their own, on the quality of the results to lead the results of scientific and technological research to the market, often opening new markets.
- A critical and practical sense of the design creativity combined with strategic skills allow the improvement of the social impact of innovations. Approaches such as scenario building and critical design not only improve communication between actors, but can also be used to evaluate and create possible *material visions* in relation to future possible and desirable scenarios.
- Strategic design skills help to create an integrated body of products, services and communication strategies, and generate and develop to create value through the use of a network of actors (companies, institutions, non-profit organizations, etc.). Indeed, the more markets are uncertain, the more companies need to broaden their horizons and become less specialized.
- Story-telling skills make it possible to communicate the meaning of a new product, material or new technology to users. These skills can translate the value of the use of innovation by communicating the benefits that the product brings to everyday life. The production of meanings that materialize in the products and the amplification of these meanings with the methods of visual communication facilitate the market's understanding of the products, thus contributing to the company's acceptance of the technical-functional innovations.
- In processes involving potential consumers or actual users, the *user experience* approach, through direct contact, interaction and the use of ethnographic

methods and observations, enables designers to understand consumers' needs and unspoken desires, and to generate benefits for both them and for companies. Designers do not receive specific demands from consumers but interpret their behaviours and possible aspirations in functional and experience terms. Designers act as interpreters in a different linguistic code on the basis of an understanding and translation of social aspirations into project elements.

Design skills are particularly well-suited to encourage the generation of material visions, starting from the most disparate types of information. However, there are still many obstacles to be overcome for the integration of design skills in the search for materials. There are factors that give hope and research experiences that promote joint collaboration between designers and technologists, and even with scientists to foster the development of the most promising social and economic technologies. Designers can provide: a new perspective on the interaction between technological and design-driven innovation to define a repeatable process for conceptualizing and designing products that use unique properties of a material designed for that purpose. There are also problems to be resolved. Among them, gaps in designers' technical knowledge that limit their involvement in the understanding and application of materials and high technologies, and gaps in tools that are useful to trigger a cross-pollination of methods and practices for effective communication and collaboration of skills, to make the products satisfactory and desirable.

The Madec team of the Politecnico di Milano intends to continue actively developing design tools and methods and to promote the debate already started with other disciplines on real joint actions for material design research evolution. We are aware that the prospect of a better future can be modelled by "Emerging Design Culture(s)", according with the concept of plural entity that includes many different cultures that engage in dialogue and debate to find the solutions appropriate to current problems and challenges (Manzini, 2016). Within this framework, society in its entirety is a huge Future Building Laboratory - a laboratory that, with its many contradictions, is already issuing signs of a new emerging culture and also of a new material design culture.

References

European Commission (2010). The European Competitiveness Report 2010. Retrieved from http://bookshop.europa.eu/en/european-competitiveness-report-2010-pbNBAK10001/.

Acha, V. (2008). Open by Design: The Role of Design in Open Innovation. DIUS Research report 08 10. Retrieved from http://dera.ioe.ac.uk/8751/1/DIUS-RR-08-10.pdf.

Chesbrough, H. (2006). *Open Innovation: Researching a New Paradigm.* Oxford: Oxford University Press.

Ferrara, M. (2009). Tra Scienza e design. Un'evoluzione intelligente. In Ferrara M., Lucibello S. (eds). *Design follows materials* (pp. 28-39). Florence: Alinea.

Ferrara, M. (2010). AdvanceDesign: scenari, visioni e advanced material per un rinnovato rapporto tra design e scienza. In Celi M. (ed), *AdvanceDesign. Visioni, percorsi e strumenti per predisporsi all'innovazione continua* (pp 151-164). Milan:McGraw-Hill.

Ferrara, M (2014). 2013. L'impatto delle tecnologie sulla pratica e la cultura del design. *Op.Cit* 149-1: 81-96.

Ferrara, M. (2015). "AdvanceDesign: A Renewed Relationship Between Design And Science For the Future". In M. Celi (ed), Advanced Design Cultures, Long-Term Perspective and Continuous Innovation, Switzerland: Springer, pp. 149-169.

Krippendorff, K. (2006). *The Semantic Turn. A New Foundation for Design London*. New York: Taylor & Francis.

Lee, J. (2015). *Material Alchemy: Redefining Materiality Within The 21st Century*. Amsterdam: BIS.

Maldonado T. (1976), *Disegno industriale un Riesame*, Milan: Feltrinelli.

Manzini, E. (2016). Design Culture and Dialogic Design. *Design Issues*, 32-1: 52-59.

Powell, W.W., Koput, K., Smith-Doerr, L. (1996). Interorganizational collaboration and the locus of innovation: Networks of learning in biotechnology. Administrative Science Quarterly, 41: 116-145.

Razzaque, M. A., Dobson, S., Delaney, K. (2013). Augmented materials: spatially embodied sensor networks. *International Journal of Communication Networks and Distributed Systems*, 4-11: 347-477.

Ribul, M. (2013). Receipts for materials activism. Retrieved from https://issuu.com/miriamribul/docs/miriam_ribul_recipes_for_material_activism/ch1j.

Verganti, R., Dell'Era, (2014), Design-Driven Innovation: meaning as a source of innovation. In Dodgson, M., Gann, D. & Philips, N. The Oxford Handbook of Innovation Managemen (pp 139-162). Oxford: Oxford University Press.

DESIGN-DRIVEN MATERIAL INNOVATION METHODOLOGY*

MARINELLA FERRARA
CHIARA LECCE

MADEC, Design Department, Politecnico di Milano

A new emerging scenario is growing within the contemporary Material Design Culture. This new scenario takes shape by growing evidence of advanced materials and performance, and the approaches to them. Advanced materials have the potential to become Key Enabling Technologies open up new possibilities for design and be crucial in the economic development. Design, as the process of "making sense of things", have been expanding its field of expertise from the products towards "ways of thinking and doing", where also materials design get its place with the aim of designing solutions for complex social and environmental situations through co-design activities (Manzini, 2016). The co-design activities have the potential of put together different innovation actors, and disciplines, each of which is the bearer of a different culture and way of thinking. In our view, diversity is not a danger to the process but a wealth. Therefore research centers, enterprises, creative communities, various experts and final users, now more than even, should be involved in the creative challenge in order to achieve disruptive innovation, both in functionality and conceptual meanings. Coupled the outside-in and inside-out process working in alliances with complementary partners in which give and take, taking also into account consumers and final users is crucial for innovation success. To shape a new future, it is necessary to combat the rigidity of traditional productive sectors and the traditional closure of disciplines and capabilities and to be able to guide and manage a systemic innovation.

Facing this new scenario, Material Design Culture Research Center (Madec) of Politecnico di Milano had been committed, over the past two years, "on developing a new

*
This essay is an updated and wider version
of some contents first published in:
Marinella Ferrara and Chiara Lecce,
"The Design-driven Material Innovation Methodology".
In proceedings of *SYSTEMS & DESIGN. Beyond Processes and Thinking 2016*
congress organized in Valencia the last year (22-24 June 2016).

specific methodology able to manage the whole design process from tailor-made material to product systems, integrating different actors of innovation, enhancing capabilities of open explorations, and reducing the time-to-market for materials and products." (Ferrara & Lecce 2016, p. 431)

On the basis of the contemporary Material Design Culture, that encompasses the design arena reported in the first part of this book, and the whole of knowledge, values, visions and qualities criteria as discussed in this last part of the book, the Madec team started analyzing theory and method, to fund elements useful to overcame the consolidated situation of design practices in materials sector, introducing new insight of the complex and opportunities that enterprises are facing, and get transparent and self-evident processes that can be easily optimized.

The result has been the Design-driven Material Innovation Methodology (DdMIM), the model that allows the development of one or more materials starting from scientific discoveries, technologies, material patents or production processes, in order to individuate new visions and applicative scenarios, to profile products lines, to develop specific products and to valorize them for the market launch. The DdMIM would be a systematic approach and an Open Innovation tool for research centers, design schools, practitioners, start-ups, and SMEs.

Due to the contemporary interest in multiple participant co-creation and collaboration, behaviors have become critically important. For this reason, behaviors synchronization is a key ingredient in any robust methodology-based and innovation-culture-building initiative (Van Pattern & Pastor, 2013). So, what the method really tries to do is to let designers, enterprises, and innovators understand how to deal with materials and materiality, contextualizing them within a wider socio-cultural scenario.

The design-driven innovation approach – operating in interdisciplinary fields – assigns a strategic role to design in R&D work, putting technical-scientific research on a par with design methods and the analysis of a material's technical-performance potential on a par with the analysis of its aesthetic and visual perceptive characteristics. Therefore the human and communities focus of design approach prioritizes the analysis of human behaviors, pleasures, and aspirations helping to understand and interpret society's needs with creativity, stimulating the creation of specific, context-oriented 'material vision' that establish a framework of innovation-oriented meanings. Designers do so while using their own tools linked to creativity, intuitiveness, and multimodality, and to the ability to shape and pre-shape.

1. METHOD'S REFEREES

As a first step, it was essential to have a wider view on the actual panorama about generic creativity-driven methodologies.

An important referee to develop a correct methodology process was the publication titled *Innovation Methods Mapping: de-mystifying 80+ years of innovation process design* edited by GK VanPatter and Elizabeth Pastor with the Humantific Lab. (2013) Essentially the publication concerns an in deep analysis of a wide variety of innovation process models created since the 1920s. In particular two main innovation process models were compared: *Applied Creativity* (or *Creative Problem-Solving* - CPS) and *Design* (or *Design Thinking*). The research found that there are three basic types of innovation process models: 'Script models' that prescribe a series of detailed actions or behaviors, often with the caveat that nonlinearity is intended; 'Zone models' that are more like scaffolds or frameworks inside which many action options are possible, often without any behavior prescribed; and 'Script/Zone models' which combines the two. Another useful indication is that most CPS process models contain graphically-depicted behaviors signals: diverge <, converge > and deferral of judgment. Also, *Open innovation*, defined as multiple, internal and external humans engaging together to address challenges with open tools, has been part applied creativity (CPS) history since the 1940s. (Van Pattern & Pastor, 2013)

The results of the entire work lead to the *Integrative Thinking*, which is the disciplined ability to recognize, orchestrating and integrating the diverse brainpower of cross-disciplinary teams as they grapple with and navigate complex innovation challenges. Integrative Thinking is about recognizing and respecting the default thinking preferences of individuals regardless of discipline, and how those preferences map to the innovation process. (Van Pattern, 2013)

A second step to define our methodology was looking outside of the design world. The D&R management is now very interested in the successful practices of 'design-driven innovation' in various industries focused on stressing Design, instead of Technology, in their innovation process (Utterback et al., 2006). To develop design-driven products companies need researchers who "envision and investigate new product meanings through a broader, in-depth exploration of the evolution of society, culture, and technology acting as interpreters who are able to envision how people could give meaning to things through intense involvement in the design discourse" (Verganti, 2009).

A research conducted in creative industries on the basis of Verganti's theory affirms: "To design new product meaning for new costumers, the company should sense the trend forecasting data which are collected with various methods by which various possible new meanings are produced. Then, the designer with his/her design paradigm helps the company do the *sensemaking* process in which the one of the possible new meanings is considered to be the best in anticipating a new trend is selected and defined. To translate the defined new meaning into a new product, the most suitable product language to express the defined new meaning –supported by selected appropriate technologies – is selected. The result will be used as the specification to develop the new product." (Kembaren et al., 2014)

Mixing and connecting all the referees previously displayed, MADEC's 'Design-driven Material Innovation Methodology' [Fig. 1] is based on a 'Script/Zone' innovation process model. It consists in the reversal of the traditional problem-solving approach to material design innovation: material doesn't exist in its peculiarity before to be chosen, but a material vision born out of the interpretation of the technical opportunity, and the social necessities coming from a community of actors that acts in order to define and develop the material and product innovation simultaneously.

Another relevant character is the 'meta-method' model because to enable multiple participant orchestrations is much more important than any technique. Today most forms of cross-disciplinary work not only require an externalized meta-framework, but deep knowledge regarding how to apply it in various innovation contexts.

2. DdMIM

The DdMIM is based on design-driven approach. Compared to the traditional R&D in the materials sectors, the Madec' methodology anticipates the contribution of design research already in the early stages of R&D, considering design as one of fundamental driver of innovation, due to its human-centered approach, capability to active solution-oriented discourse, and envision new scenarios of applications, involving a variety of actors and co-design with them. Design approach focuses on knowledge, values and quality criteria that emerge from the interaction of the variety of innovation actors during research activities.

Considering emerging materials and technologies to be developed and applied, during the process the method puts a strategic focus to their applications that could satisfy people's needs and aspirations in specific socio-cultural contexts considering also its perspective of changing. This approach puts a strategic focus to the user's experience and the meaning that a material assumes. During the process designer is called to understand materials, technology opportunities and their cycles of life, to interpret what could be meaningful to people in order to offer new solutions and new meaning that are intended to be applied in products.

The design can derive from different start points, that defines the challenge:
- A new material proposal to be developed on the basis of exploratory samples experimented with a low-tech approach, like experimental material come from a DIY perspective;
- A novel semi-developed material came from scientific laboratory;
- Fully developed materials not very well-known and exploited in its application areas;
- An existing and traditional material, that look for new performances and identity;

Since new or old materials need to be related to using meanings that change, the design process can help to selected application area and interpret their features and identities, or updates expressive language responding to existing trends.

During the research process, the characteristics of materials are analysed, defined, designed and explained from several points of view: the performances, that can be easily translated in user experience; the sensory characteristics, that can be easily translated into analogous, measurable physical properties (technical properties); aesthetics languages elements, which express the relationship between the product and its intangible meaning; the cultural perception that may relate to prejudices or anxieties related to materials in specific contexts of use.

The process can involve, in addition to materials science, engineering and biological skills, also ethnoanthropology, pragmatic aesthetics, and sociology, that allow expanding the 'design discourse' to understand the change in society and to draw the future perspectives for material and product design. These skills are applied in different phases: the analysis, design, prototype, assessment and storytelling phases to understand the reaction of potential costumers about functions, behaviors surfaces, textures and aesthetics of a material and their applications. Sensoaesthetic, somaesthetics and perception studies, as well as sensemaking, help to envision storytelling, considering these entire fundamental to a customer consensus to a new material. This cross-disciplinary process integrates meta-design, material design and product design steps with the aim of orienting technological researches towards social innovation meanings.

The DdMIM give shape to a systematic approach to R&D. DdMIM is made up of 6 different steps. It starts with an open challenge by a techno-scientific opportunity in the field of materials, as described above, i.e. in other words, scientific discoveries, a patent, a new production process or a material problem to solve.

The first steps are "data collection" and "sensing", where the researchers analyze the initial data (technical constraints and marketing opportunities) to understand the background. Then, in the sensemaking phase, they start to envision applications scenarios, to outline a 'vision of matter' that allows defining the user experience, on which profiling materials and product lines design. The sensemaking phase converges to product concepts, to defining material qualities, performance, and behaviors to be developed on the basis of technological platforms (specifying phase). Then a "design discourse" contributes to defining the design process. Once material and products designs are defined, it's time to go into the prototyping step, followed by the storytelling step, in order to make clear the message of innovation for customers ('setting up' phase). Finally, after an assessment step, the design process is finalized with its strategy, which contributes to material positioning inside the market.

The DdMIM is capable of leading to radical change that is recognizable and replicable.

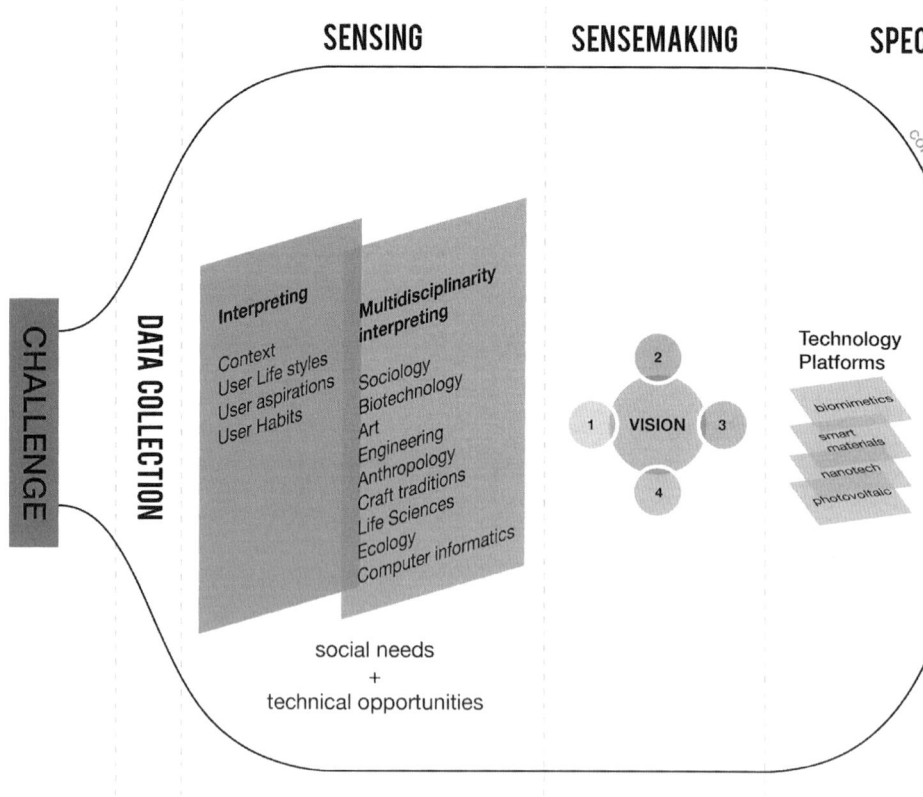

3. DDMIM PHASES

The challenge of the DdMIM is based on the different four start point, above described, and on the modality of the project born. The project can start with a whole group of partners (enterprises, consultants, and research centers) involved in the DdMI process, or with a little group of actors that can be extended during the process, to be involved in different stages. This community-based participatory research model require to individuate, select and engage stakeholders as a partners, who impact, and are impacted by a material and manifacturing opportunity or problem of concern. The collaborative approach to research might equitably involve all partners in the research process, and recognizes the unique strengths that each brings. Among the phase of the process, the 'specifying phase' is a topical stage were the team of collaborators and participants could become wider.

Fig. 1 The Design-driven Material Innovation Methodology by MADEC.

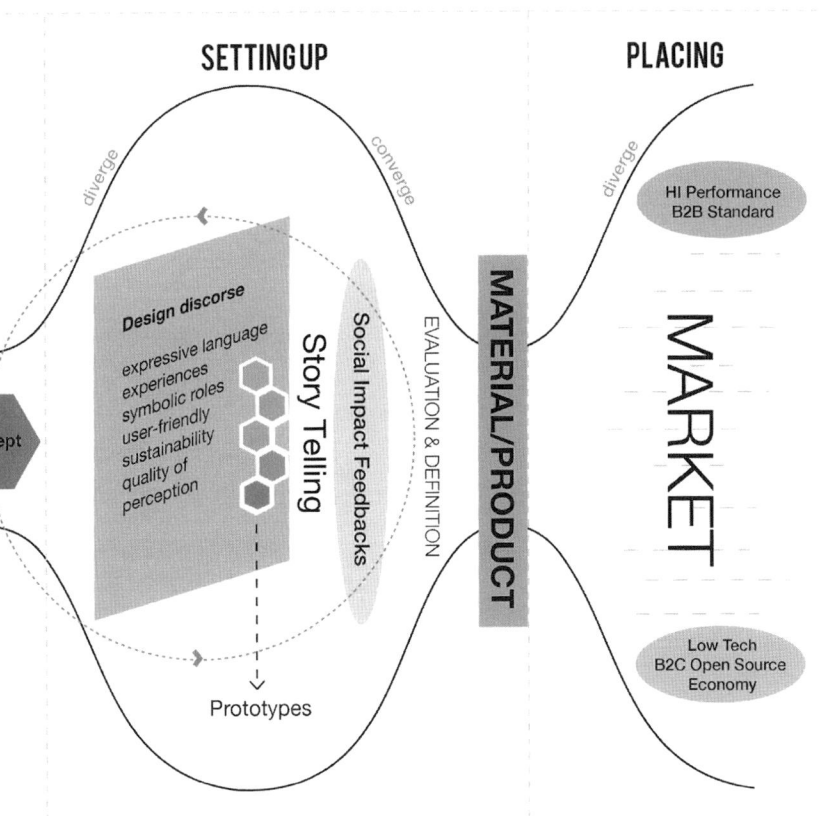

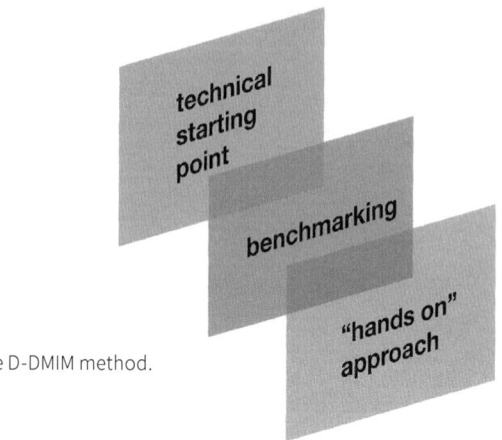

Fig. 2 Data Collection phase from the D-DMIM method.

3.1. Data Collection
Data Collection is the first step of DdMIM. It is a 'technical-cognitive' and 'sensory-analytical' phase about materials and its characteristics, that use both quantitative and qualitative approaches.
Starting with the technical material characterization of a raw or semi-developed material, or also a full developed material that wants chance, the process require to analyze the material itself and its options. This step includes three main actions [Fig 2]:
- A deep analysis of the starting point material its production technique and cycle of life in order to acquire technical knowledge from scientists, technologist, and material suppliers. This allows to acquire information about material potentialities, its opportunities to be molded and manipulated gaining sensorial characteristics defining the material for 'what it is', 'what it can do', 'how it appears' and its behavior. It could be also useful to compare the material with similar o alternative ones to highlight differences and similitudes;
- Activities of benchmarking positioning the material in the contemporary materials scenario, among similar or different, without any preclusion to other production sectors, in order to find space for opportunity. This is a way of discovering which could be the best performance to achieve in the application, use – whether in a particular sector and in many other sectors. The information can be used to identify gaps in production in order to define potential spaces and new market;
- To 'feel' the material trough manipulation: 'hands on' approach and interpret the sensory potential of the material. About this last point it is important to stress on how physical encounters with materials or the aesthetic experiences that derive from hands-on manipulation of materials . This can positively influence the creative process. (Nimkulrat, 2012)

3.2. Sensing
The term 'sensing' means the perception that something has occurred or some state exists. Designers are able to scan lifestyles in a certain socio-cultural context, to under-

stand people's needs, their motivations and capabilities, and the social dynamics in which they are living. They recognize users tastes orientation and interpret them. This ability is fully applied in sensing phase in order to situate the chooses while bringing alternative contributions and stimuli as well as introducing new paths, parallel visions, and uncustomary detours to the research done by chemists and engineers.

Sensing activity is very important to fix the data and the project context features (economic and socio-cultural specific situations, etc.) as well as cultural and behavioral characters of society in changing. This phase could involve different experts of various disciplines: humanities as sociologists, anthropologists and ethnoanthropologists, psychologists and pragmatic aesthetics with their tools of analyses, as well as life sciences and also users, in different section of analytical work, in order to expand the 'design discourse' to understand the change in society and to draw the future perspectives for material design [Fig. 3].

This stage is useful to 'extrapolate' information for the sensemaking stage as users' attitudes, necessities, and aspirations in relation to their physical, physiological, ideological and social well-being, as emerging trends elements. Sensing activity let decide what kind of culture you want to create in the role of innovation process organization.

3.3. Sensemaking

In design language, the 'sensemaking' is the activity of developing ideas about what could or should be possible and desirables, overcoming the present situation. In a common design process, sensemaking has a critical role in bridging the gap between

Fig. 3 Sensing phase from the D-DMIM method.

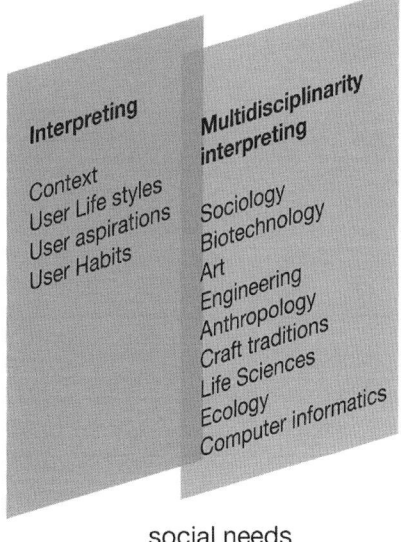

the brief, the ideation, and conceptualisation stage. Sensemaking can be both a personal internal and reflective activity, informed by the individual's unique perspectives and frames; and, an external collaborative activity resulting in a pooling of ideas informed by meanings made by different individuals. But, as suggested by Van Pattern and Pastor (2013), to use only data, in shape of words and numbers, to explain new ideas, rather than drawings and pictures, joint to words and numbers, improves understanding during the process and its outcomes. So the designer role in the sensemaking is relevant, due to designers are familiar with the act of sensemaking, visual representation, and communication of ideas.

In the book, *Abductive Thinking and Sensemaking: The Drivers of Design*, Jon Kolko describes sensemaking as, "making sense of a situation taking into account unique experiences and discerning connections."

In our method, sensemaking is a design activity related to interpreting and communicating a new scenario of materials and applications coming from the analysis and perception of the data, needs and socio-cultural values in the context of reference. Sensemaking leads to tangible proposal solutions, which give meaning to the human experience. The problems or aspirations setting belongs to the mechanisms that we start to make sense. A well-defined problem, in its form and substance, transforms nebulous and meaningless discomfort into a structure full of meaning. So without understanding the situation of start, the context being designed for, the sensemaking phase fail.

In the DdMIM, the sensemaking phase gets to a 'material vision', that express the meaning of the changing and what it bring in the human experience with the materials in its application [Fig. 4]. The material vision makes tangible what the new material and its applications represent for users experience, anticipating new trends.

At this stage can be useful take as reference points for the generation of new ideas, on the basis of the emerging socio-cultural trend, the emotional aspects while interacting with products. According to the Jordan model (2000), derived by Tiger (1992), four types of human pleasure can be distinguished: physio-pleasure, socio-pleasure, psycho-pleasure, and ideo-pleasure.

Physio-pleasure represents pleasure experienced through the senses, such as taste, smell, and touch. Socio-pleasure is defined as the type of pleasure that arises when we are socially interacting with friends, colleagues, and people we love. Psycho-pleasure

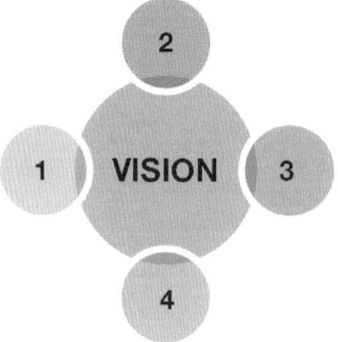

Fig 4. Material Visions by the Sensemaking phase.

refers to emotional or cognitive satisfaction. Ideo-pleasure refers to ideals culturally, aesthetically, by appealing to the consumer's values.

In the sensemaking stage, designers have to use imagination to open up different possibilities defining a specific material vision where the general concept can be collocated. The most relevant research questions focus on the 'Why'. Following the why the research should reply to other tree questions: 'What if', 'How', and 'What'. So doing from a material vision can result in a number of materials, product or service concepts to be developed and prototype. The result of this process will be used in the specification to develop a system of new concepts. Through generating more product concepts, the material vision strengthens its meaning.

3.4. Specifying

The result of sensemaking will be used in the specifying phase that allows completing the information needed for define the materials requirements, to be proposed to materials scientists, considering the areas of application suitable for the exploitation of the material vision. The specifying stage is one of the most complex and research intensive portions of the process that leads to the performance definition of material variations. We speak the plural because as argued by Norman & Verganti "By forcing the design team to simultaneously diverge into multiple directions, this enhances the chance of having some of these attempts start off in a different design space, one that might possibly allow for a successful, novel new product." (Norman & Verganti, 2012) This statement is particularly suitable for material development. While specifying the solution domain of available technological process, here named technology platforms, the process focuses on capturing the performances of the solutions provided by the innovation process partners.

In that stage design, materials scientists and engineering have to work strictly to take the right decisions to go to into the stage of the experimental prototypes of the new material variations. To choose the right technology platform, taking advantage of the multidisciplinary approach, is one of the most important aspects of ensuring a project's success.

To keep up with the complexity of modern industrial production, multiple directions can be taken into consideration for obtaining a performance in order to conceive applications in which the material expresses itself and can be exploited as effectively as possible. The specification traits include respecting the laws that guarantee security in the areas of applications, as well as environmental aspects, and such technical performance features as durability, processability, and mechanical properties, insulation and barrier; also included are such perceptive traits as the material appearance to be developed in the next phase. A material's composition and the results of perceptive variations mainly produced by processing procedures are both taken into consideration. Technology platform and processing procedures that can be compared in order to scientifically guide the material selection process depending on the product concepts we intend to introduce to the market. In this way, we can avoid merely providing a list of possibilities from which to choose subjectively and seek that aspect of envisioning that is typical of design right from the start, which is thus positioned at the very heart of the scientific process of material production. Last but not least, 'palettes' or 'scales' of materials function as a warehouse/archive of scientific importance for future possible uses of the materials developed, makes way for application processes typical of product design, rather than restricting them to one single final choice of a market-ready product.

This process converges to the 'product concept stage'. This defines the idea of the material, of its performance and behaviors on the base of scientific advances and technological platforms [Fig. 5].

3.5. Setting up

When the product concepts, with the quality of the material, is defined we can start a development stage of the material applications in products design. By analogy to what Kembaren argued: "To translate the defined new meaning into a new product, the most suitable product language to express the defined new meaning –supported by selected appropriate technologies – is selected." (Kembaren et al., 2014), this is the time of come back to the "meaning" and its relationships with such cultural aspects as a setting's productive human capital, and territorial assets, to recover new insight or to go in deep in the meaning, founding expressive and communicative opportunities offered by the innovation of materials and technologies. The material identity is to be expressed and understood as a complex of opportunities, but also of limits and weaknesses, that characterize it and that must be elaborated, embedded in a shape, in order to conceive applications in which the material expresses its meaning. The identifying traits include perceptive quality, like the ability to transmit or to reflect light, colour, and other visual elements of appearance, as well as tactile and sound characteristics such as features as well. Sensoaesthetic, somaesthetics and perception studies, fundamental to a customer consensus to a new material, are called to go deeper, in order to express innovation value through a linguistic adequacy and make it more relevant to users' needs, according to consumer values, thus more easily accepted. That's why, during this phase the characteristics of materials are analysed from several points of view: the sensory experience, that can be easily translated into analogous, measurable physical properties (technical properties); aesthetics languages elements,

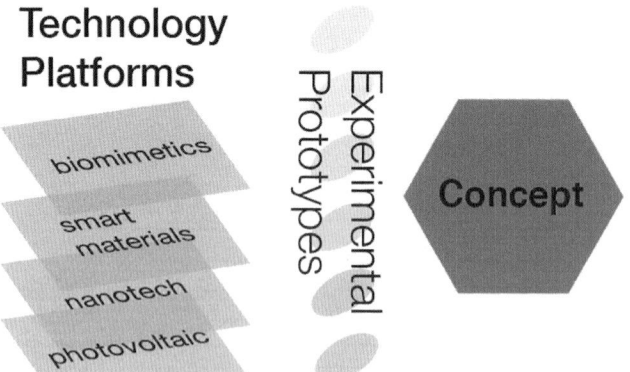

Fig. 5 Specifying phase from the D-DMIM method.

which express the relationship between the product and its intangible meaning; the cultural perception that may relate to prejudices or anxieties related to materials in specific contexts of use.

At this stage to evaluate the linguistic adequacy the process could utilize open tools and approach with the participation of potential user's called to evaluate the appeal and interest of the various proposals developed. So it is possible to understand the reaction of potential costumers about functions, behaviours surfaces, textures and aesthetics of a material and their applications.

A really strategic step of this phase is the 'storytelling'. In fact, to deliver the new product with its defined new meaning into the markets, a storytelling has to be carefully designed along with its product language to amplify and to relate the message of the defined new meaning to the mind of the potential customers.

The role of design is thus to look more comprehensively at the new product, to respect the technical insight for considering sensory, emotional and symbolic qualities and, from this, giving a message to the consumers: a meaning that increases the value of products in the market. Also, the 'user acceptance' of a material/product depends on the correspondence of the experiences with the society needs or cultural trends, as well on the ability to communicate the meaning of material/product innovation. Visual communication skills are fundamental in this process: the message should be pleasant and easily understood by consumers to let the new product be accepted, desired and chosen by the consumers.

The design discourse, using its own linguistic tools must be able to keep up with the contemporary languages and its relationships with such cultural aspects as human capital, and cultural territorial assets. When the design discourse on the material qualities is defined it can start a development stage of the material applications for products design.

Finally, organizing a material experience session of potential users the innovation social impact could be evaluated before to place the product on the market [Fig. 6].

3.6. Placing

In the end, design can define how to place the material/product on the market. In this phase, it is suggested to consider at the same time various distribution channels to the ultimate purchaser or end-user.

Today it is important to assure that the product position inside the market diverges from different possible approaches connected to production processes – Business to Business (B2B) or Business to Consumer (B2C). But getting closer to this phase becomes competence of business models experts with the help of designers. The design contributes in this stage can be mostly the implementation of the visual communication of the product [Fig. 7].

4. SOME CASE STUDIES

To better make you understand the various steps that the methodology articulate, and the possible output, we propose below two case studies born as a creative start-up enterprise that based on a design approach, launched a new material developed in a system of different products and service.

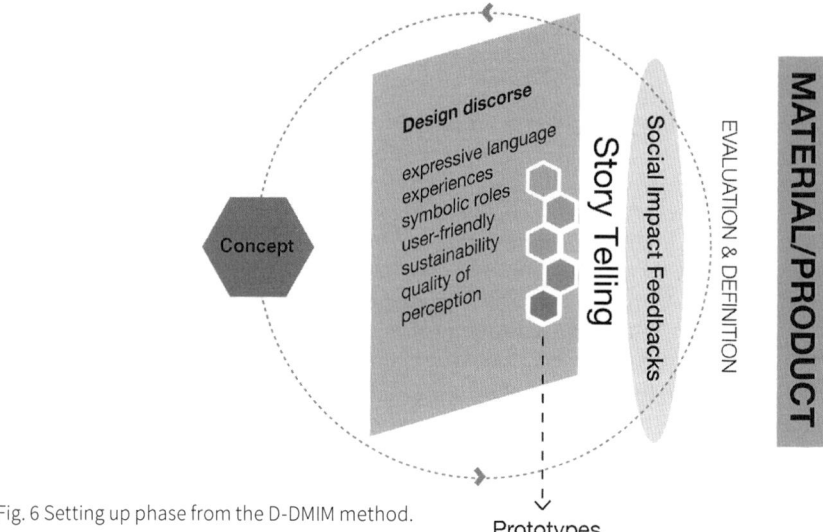

Fig. 6 Setting up phase from the D-DMIM method.

Fig 7. Placing the final phase of the D-DMIM method.

The first case study is the Wood-Skin®, an Italian creative start-up enterprise by a talented team of four members with a different background, Milan-based and working together. The variety of skills and approach brought by each member of the team make possible to convert a design proposal of a new material into an enterprise that produces and distribute a variety of materials wood-based, offering practical applicative solutions and services in the area of interior design. The vocation and area of knowledge of each team member have merged seamlessly to produce a well-rounded group, each complimenting the others. Applying a design approach, the company has been developing a material innovation coming from an old and traditional material, like the wood.

The brand developed and patented a technology that has been redefining wood properties and generate a number of materials/products by a material vision of "a folding wood that can apply to cover every element of architecture giving a sculptural appearance". Their vision derives from a "specifying" phase that combines together different characteristics starting from a traditional material revolutionizing it through transforming it into a composite material and a new one-step fabrication process.

Wood-skin, combining the rigidity of traditional materials with the flexibility of textiles, it allows countless applications both to entirely customizable architectural and design elements. The material tissue-alike is able to be shaped in many forms to create flexible and customized surfaces, yet retaining their structural and aesthetic value intact, simultaneously. More over Wood-Skin is also a wide collection of different textures and composites produce based on the same material vision and process, and related to a concept exploited in different product with different properties (high acoustics, light weight, fire proof rated, etc.) futures (durability, adaptability, colours and finishes selection, many different patterns) environmental aspects and values (sustainability of the process and material, dry assembly and easily removable, flat shipping, sculptural effect, integrable with light). Through their design approach, the team has achieved to be always ready to face every new challenge and to apply their individual skills to offer a design and consulting activity, in addition to the product itself.

The second case study is Litracon®, particularly choose as case in point about storytelling of a material. Litracon is a translucent concrete (the name is short for 'light-trasmitting concrete') designed and patented by the Hungarian architect Áron Losonczi, working with scientists at the Technical University of Budapest, who tackled the issue of glass in architecture, learned about optical fibers. During is low-tech experimentation with glass fibers cast in the concrete mix has given the materials a new dimension, a new material vision of a 'concrete that light up passing light trought' and experimental prototypes. In order to advance his experiment (2001-2003), Losonczi made a partnership with Schott, a world's leading manufacturer of optical fibers. In 2004 the experiments finished and the company was founded to produce the first type of this special material, made of 96% concrete and 4% by weight of optical fibers. After the first result, Litracon Classic®, the material research has gone ahead to make the process easier and industrialize and to be the material good to be used in a variety of way and desired shapes. Only 1 year after Litracon pXL® was patented. Unlike the first product, the new pXL® use a specially formed and patented plastic unit, easy to introduce in array into the concrete. This and the industrialized way of manufacturing bring the new pXL® material into a more affordable price range. So the way of manufacturing

Fig. 8. Wood-Skin®'s concept proprieties.

is more industrial, the cost is reduced and the visual appearance of the material change. It is more uniform, as the light dots appear with regular distribution on the surface. It is easy to create patterns or even coloured logos out of the pixels.
For its aesthetic value Litracon was used in furniture sector with Litracube lamp, and in purpose for jewellery in a new appearance with optical fibres in organic distribution, commercialized by the Litracon company. Contrary to Litracon Classic®, in Litracon Classic® was listed among the most important inventions of the 2004 by TIME magazine but to get people meet and confident with this new material/product the company understood that it was necessary to communicate a lot about it. So 'You Are Energy' is the name of the installation designed by Gagarin Ltd. with Tvíhorf Architects in Island to catch people's attention. The installation was higly interactive. It consisted of a big, interactive concrete wall made of Litracon Classic® blocks which illuminates from within when a force is applied to it. Visitors were invited to test their strength and met the challenge of inducing an explosion. During the exhibition they can learn about the material, and its environment impact. It is a solid material stronger than glass in order to be apply in architecture. When a solid wall has the ability to transmit light, it means that one can use fewer lights during daylight hours. Other advantages in addition to less energy consumption, are the possibility to illuminate pavements, the finishing surface, and routine maintenance not required. The material has succeeded in the market for its distinctiveness.

These two cases show that a material innovation success depends on two other relevant design outcomes: distinctiveness and user acceptance. In a world were the number of products becomes greater and greater, material can make the difference to the competitors, for distinctiveness of your products in the market. The material innovation value and message should be pleasant and easily understood by consumers, that means to increase the ability to communicate the material innovation.

5. CONCLUSIONS

The DdMIM help to deals with the complexity of a product/material R&D open innovation process were many and very different actors are involved. Here we summarize the principal DdMIM futures:
- It is based on material design and product design integration, trough a deeper understanding of material qualities;
- It focuses on new values and meanings, centred on human pleasure and consumers needs;
- It enables the collaboration between researchers, designers, and companies through the open innovation processes;
- It integrates multi-skills in a cross-fertilization process;
- It is possible to create a different teamwork for each phase of the process;
- It is useful for envisioning and developing new product concepts;
- It is a non-linear obligatory sequence with a start/end process, indeed it is also possible to begin the procedure from an intermediate step of the methodology.

Actually the DdMIM is part of the 'Design for Enterprises' the European training program for SMEs, started this year and operating for the next two years in order to encourage SMEs to use design as a tool for their competitive edge. In the Design for Enterprises training program the DdMIM is part of the module "Design for Materials", and help enterprises to manage a materials design process for product and services innovation where different actors like materials scientists, suppliers, creative communities and consumers are getting engaged.

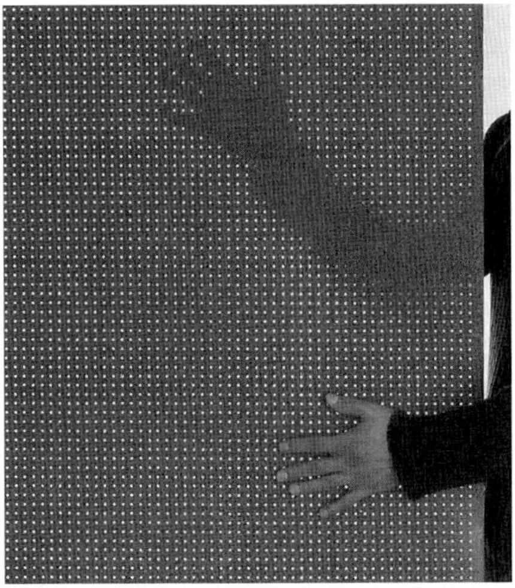

Fig. 9 Litracon's concept proprieties.

References

Caisse, S. & Montreuil, B. (2014). Polar Business Design. *SAGE Open January-March 2014*. [Consulted: 4th of November 2015] http://sgo.sagepub.com/content/4/1/2158244014522632.full.pdf+html.

Ferrara, M. & Lecce, C. (2016). "The Design-driven Material Innovation Methodology". In *SYSTEMS & DESIGN. Beyond Processes and Thinking 2016*, Proceedings June 22-24, 2016. Valencia: Editorial Univertitat Politècnica de València. pp.431-448.

Jordan, P.W. (2000). *Designing Pleasurable Products* Philadelphia PA: Taylor & Francis.

Kembaren, P., Simatupang, T. M., Larso, D., Wiyancoko, D. (2014). "Design Driven Innovation Practices in Design-preneur led Creative Industry" in *Journal of Technology Management & Innovation* (9) 3, Santiago.

Kolko, J. (2010). Abductive Thinking and Sensemaking: The Drivers of Design. Design Issues, 26,1: 15-28.

Manzini, E. (2016). Design Culture and Dialogic Design. Design Issues, 32,1: 52-59.

Tiger, L. (1992). *The Pursuit of Pleasure*. New York: Little Brown & Co.

Utterback, J. M., Vedin, B., Alvarez, E., Ekman, Walsh Sanderson, S., Tether, B., Vanpattern, Gk. & Pastor, E. (2013). *Innovation Methods Mapping: de-mystifying 80+ years of innovation process design*. OPEN Innovation Consortium.

Vanpatter, Gk. (2013). *Humantific Innovation Methods Mapping*. Systemic Design Conference, Oslo School of Architecture & Design, Oslo Norway.

Verganti, R. (2009). *Design Driven Innovation: Changing the Rules of Competition by Radically Innovating What Things Mean*. Harvard Business Press.

Verganti, R. (2006). *Design-inspired Innovation*. World Scientific Publishing Company.

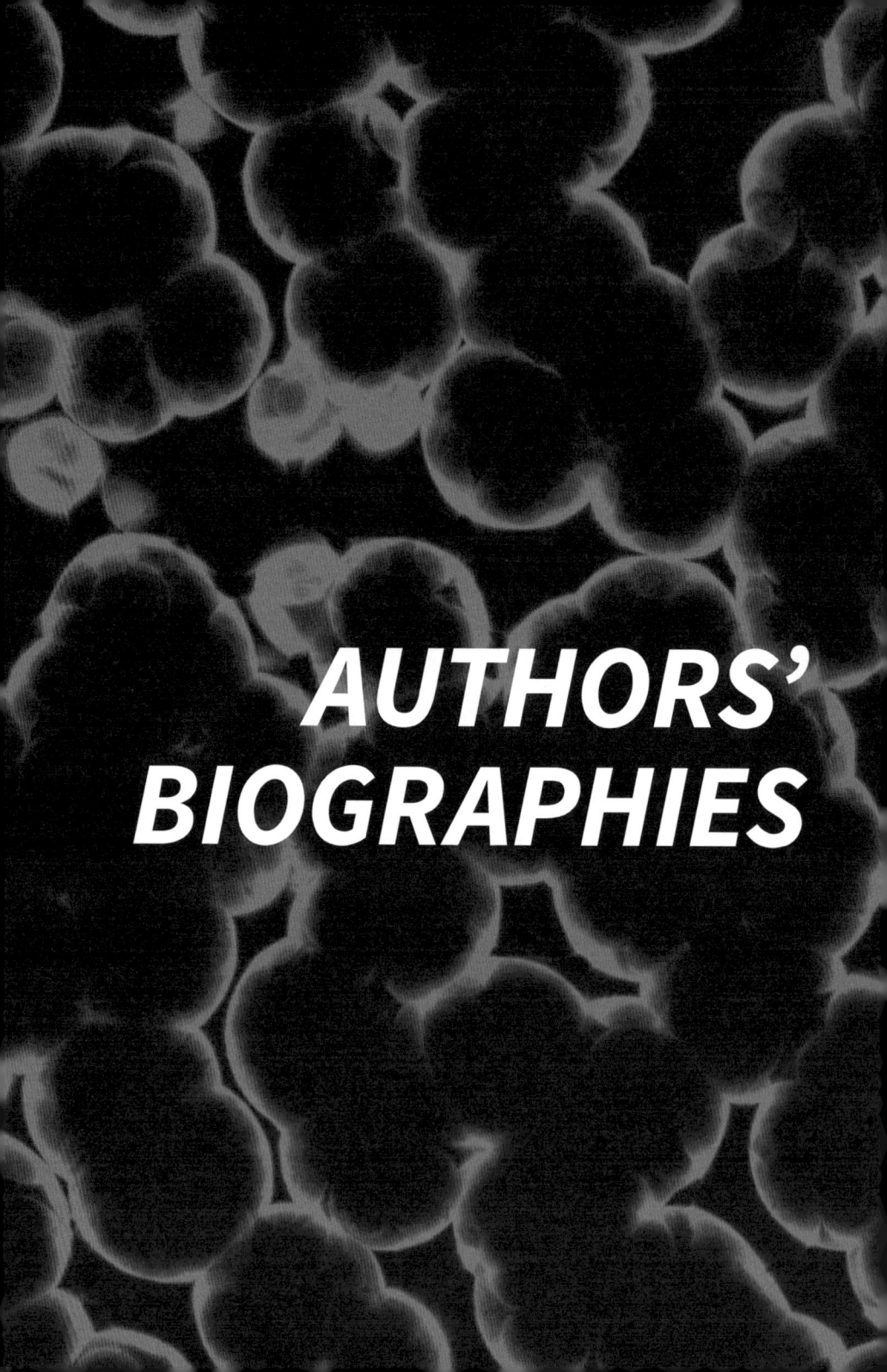

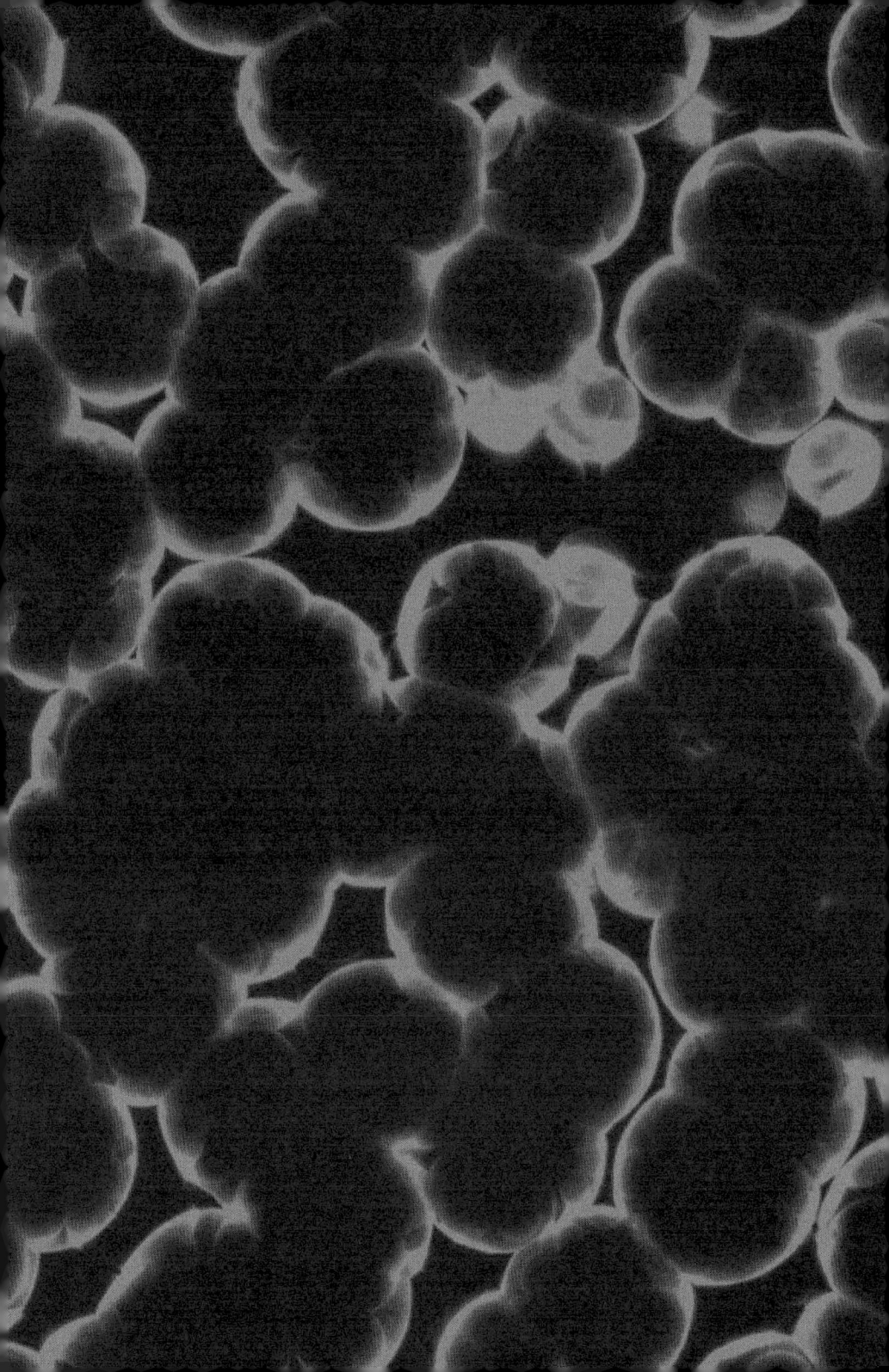

Marc Augé is one of the most influential anthropologist of the contemporary age. He is a French Ethnologist and anthropologist, he is renewed for his studies on contemporary worlds and his attention to the global and cosmopolite dimension that associate colonial communities and the Western world. He has been director of the École *des hautes* études of Paris. He elaborated the theory of "non-places" to refer to places of transience and de-culturalized that do not hold enough significance to be regarded as "places". Theory expressed in his most relevant publications: *Un ethnologue dans le métro* (1985) and *Non-lieux: introduction a une anthropologie de la surmodernité* (1992). He has recently published *L'anthropologue et le monde global* (2013), about the paradoxical effects of globalization.

Giampiero Bosoni, Full professor of Interior Architecture and Exhibition Design at the School of Design of Politecnico di Milano. Architect and expert of Design History and Theory, he collaborated with Figini and Pollini architects, Vittorio Gregotti and Enzo Mari. He has realized several articles and collaborated with specialized journals and magazines: *Domus, Lotus, Abitare, Casabella, Interni, Ottagono, Crossing, Progex* (director), *Print, Rassegna* (editor), *Brutus Casa, Pluriverso* and others. He curated the exhibitions: "Museo del Design", giving birth to the first core of the historic collection of Italian Design of the Triennale di Milano; "Made in Cassina" (Triennale di Milano, Tokyo, Paris and New York); "Il modo italiano. Italian design and Avant-garde in the 20th Century" (MBAM Montreal, ROM Toronto, MART, Rovereto). He is author of several books for Skira and MoMa among others.

Clino Trini Castelli, designer and design theorist. Frome the 70s he is pioneer of the emotional development of products identity and of the esthetic sustainability of the industrial sector. With Castelli Design studio he is involved into international activity concerning product design, strategic design and CMF planning and design. He is author of several publications and books about these topics and he perceives teaching activities by the main international schools of design. His work has been award with European, American and Japanese rewards, including ADI Compasso d'oro, the IBD Gold Award and IF Product Design Gold Award.

Giulio Ceppi, architect, designer and PhD. From 1990 to 1997, he was the coordinator of the Domus Academy Research Center, and from 2004 to 2007 the first director of the DA Master in Business Design. Since 1995 he is a researcher at Politecnico di Milano and a professor of Industrial Design in the Design School. His activities are focused on sensorial design, new materials and technologies, and design strategy. After the experience of senior design consultant at Philips Design (1998-2001), in 1999 he funded Total Tool, a design network (Milan, Buenos Aires and Tokyo) consulting and designing new business ideas, exhibitions. He's author of a number of books about innovation practice: Children, space, relationship (1997), Oggetti esistibili (2005), Design epigenetic (2009), Design Storytelling (2011), Awareness Design (2012).

Roberto Cingolani, physician, expert of the robotics and biomimetic. From 2005 he is Scientific Director of the IIT (Italian Institute of Technology) of Genova. He has been staff member at Max Planck Institut for Festkoerperforschung in Stuttgart (Germany), visiting Professor at the Institute of Industrial Sciences, Tokyo University (Japan), visiting Professor at Virginia Commonwealth University, Richmond (USA), professor of General Physics at the Engineering Faculty, Lecce University (Italy) and Founder and Director of the National Nanotechnology Laboratory (NNL) of INFM in Lecce (Italy). He is author of about 750 papers in international journals and he holds about 48 patent families in the fields of nanotechnology, robotics and biotechnology.

Fortunato D'Amico, graduated in Architecture, deals with transdisciplinary connections: art and architecture, philosophy and science, etc. He has been working on catalogs of various international publishers and has been producing some television broadcasts for the SKY network, such as Archibalena. He is among the organizers of the Dedalo Minosse International Award. He is the curator of the exhibitions such as: Vergilius d'Oro Prize and exhibition in Mantova, 2010; Architecture Laboratory in Milan, 2010; Nino Mustica: Pittura solida in Pietrasanta, 2010; Think Green, Milan, 2010; Chiara Dynys: Labirinti di memoria - Più luce su tutto - in Rome, 2010; Culture_Nature 2010, at Venice Architecture Biennale; Urban Solutions 2009 in Milan; Arte in Luce 2009 in Turin. He has been teaching with a temporary appointmente at the Politecnico di Torino, the Politecnico di Milano and the Accademia di Brera.

Massimo Facchinetti, is an architect and designer involved in environmental design since 1997. He has been teaching with a temporary appointment in the School of Architecture and Design of the University of Milan, Florence, Turin, Brescia and the Politecnico di Milano. In 2004 with Carlo Bono and Alessandra Boccalari, he founded Facchinetti & Partners working in industrial consulting and eco-building. He is a member of AEREC (Accademia Europea per le Relazioni Economiche e Culturali).

Marinella Ferrara, architect, PhD, associate professor of Industrial Design at Politecnico di Milano, and the coordinator of MADEC (www. madec.polimi.it). Her research and teaching interests focuses on the relationship between design and scientific/technological innovation, with a particular focus on emerging materials. In the last years she has been engaged in the EU project *Design for Enterprises*, and in the Executive Board of *ADI*, coordinating the scientific committee for long-term professional development of designers. As an expert in the communication of the design contents, from 2011 she has been the editor in chief of the scientific journal *PAD. Pages on Arts & Design* and guest editor of *AIS/Design. Storia e Ricerche*. She is the author of over 100 publications among books, essays and articles in international journals.

Chiara Lecce, MSc, PhD in Interior Architecture & Exhibition Design. Since 2008 she is involved in History of Design classes and Interior Design Studios at the Design School of the Politecnico di Milano. Since 2009 she collaborates for the Fondazione Franco Albini and with other well known Italian design archives, other than continuing to work as a freelance interior designer. From 2013 she is the managing editor of the scientific Journal PAD (Pages on Arts and Design) and member of the AIS/Design (Italian Association of Design Historians), as well as author for several scientific design Journals. She is currently research fellow and lecturer at the Design Department of the Politecnico di Milano. Since 2016 she is a tutor among the European project "Design for Enterprises".

Stefano Marzano, has been the former Chief Design Officer of Royal Philips Electronics for 20 years since 1991. He is the Founding Dean of THNK, the Amsterdam School of Creative Leadership.
Author and editor of several publications about design, till 1998 he has been professor at the Domus Academy in Milan and visiting professor from 1999 to 2001 of the Politecnico di Milano. He is member of the European Design Leadership Board. He received the honorary PhD in Design from the Sapienza University in Rome and from the Hong Kong University Polytechnic. In 2005 Businessweek nominated Marzano one of the four "Best Leaders: Innovators" in the world and in 2001 he has been awarded of the World Technology Award for Design from the World Technology Network.

Maurizio Montalti, is a designer, researcher, artist, and engineer. In 2010 he founds Officina Corpuscoli, a multidisciplinary studio based in Amsterdam inspired by living systems and organisms. The studio's work has been internationally shown in museums, exhibitions and festivals. He is the co-founder of Mycoplast, a company focused on industrial scale-up of mycelium based materials. Currently he is co-heading the MAD Master at Sandberg Instituut, holding a Research position at Design Academy Eindhoven (DAE), as well as teaching in different national and international academies and universities. He is also co-founder and forming part of the Amsterdam-based WNDRLUST collective, alongside Mike Thompson, Sonja Bäumel and Susana Cámara Leret.

Michelangelo Pistoletto, Italian painter and sculptor, relevant exponent of pop art and Arte Povera, since the Sixties he has developed original artistic solutions, experimenting several materials and techniques, in order to actively involve the public inside his opera. Among the numerous recognitions received: the Leone d'Oro for his lifetime achievements at the 50° Biennale di Venezia (2003); the Wolf Foundation Prize in Arts (2007); the Praemium Imperiale of the Japan Art Association (2013).

Francesco Samorè, PhD in business history and business finance. Its main areas of research are: the structure and protagonists of the contemporary Italian economy and the analysis of energy utilities. He is the scientific director of Fondazione Giannino Bassetti, with the mission of promoting responsibility in innovation, following the relationship with affiliated institutions, both in Italy and abroad, on the four main operating areas: Innovation and Governance Responsibility; Responsibility for Innovation and Life Sciences; Responsibility for innovation and enterprise; Responsibility for innovation and risk. He has been collaborating with the Department of History Studies of the University of Milan (since 2006) and with the Department of Sociology of the Università Cattolica (since 2008). He is a member of the Italian Society of Economic Historians (SISE). Since 2006 she has been a researcher for the Business Culture Center in Milan.

Giuseppe Testa, heads the Laboratory of Stem Cell Epigenetics at the European Institute for Oncology (IEO) in Milan, where he is also Deputy Principal Investigator in the Research Unit on Biomedical Humanities. He is the cofounder of the interdisciplinary PhD program FOLSATEC (Foundations of the Life Sciences and Their Ethical Consequences) in Milan. He accomplished his academic education in biology at the Embl of Heidelberg and in bioethics and science sociology at the Manchester University and Harvard. He published on the main journals concerning biomolecular, bioethics and sociology diciplines.

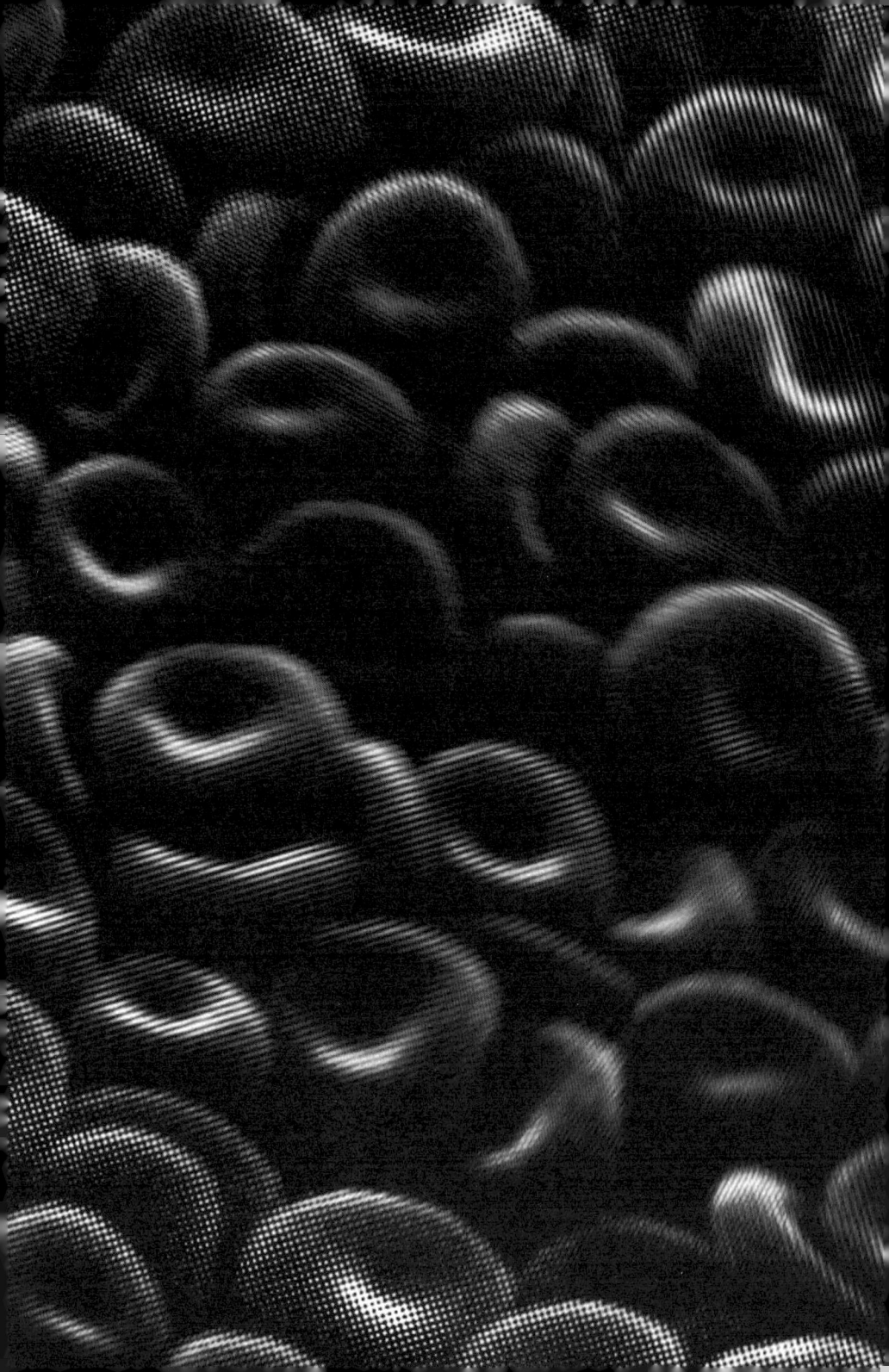

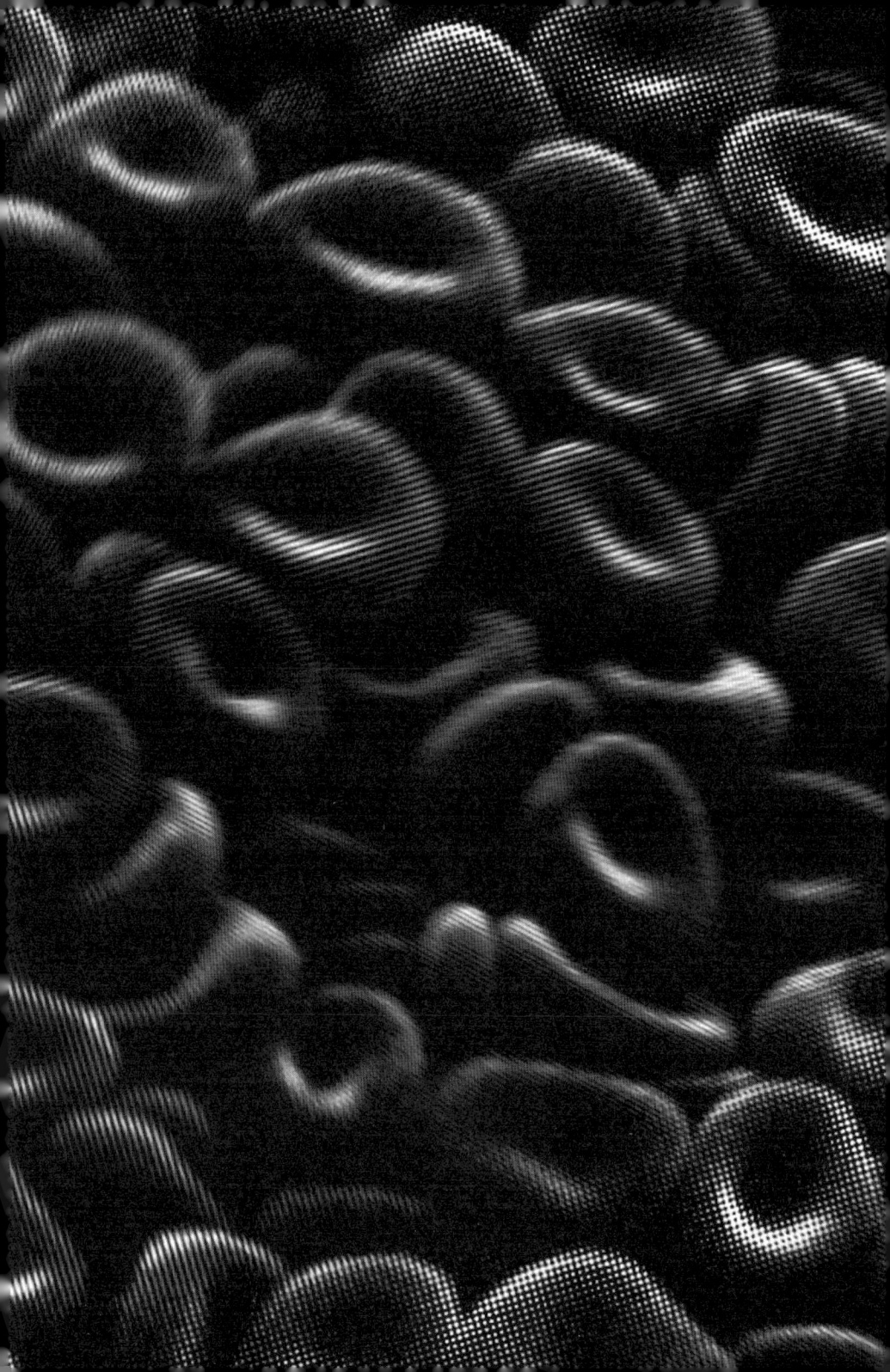

Published by
LISt Lab
info@listlab.eu
listlab.eu

Production
GreenTrenDesign Factory
info@greentrendesign.it

Editors
Marinella Ferrara
Giulio Ceppi

Editorial Director
Alessandro Franceschini

Editorial Assistant
Gioia Marana

Art Director & Graphic Design
Blacklist Creative Studio, Barcelona
blacklist-creative.com

Graphic Production
Arianna Scaglione

ISBN 9788899854553

Printed and bound in the European Union,
Settembre 2017

The manuscript of this publication has been subjected to a peer review process by international experts.

All rights reserved
© of LISt Lab edition;
© of the author's texts;
© of the author's images;

Promotion and distribution
Messaggerie Libri, Spa, Milano,
toll-free number 800.804.900
assistenza.ordini@meli.it;
amministrazione.vendite@meli.it

International Promotion and Distribution
ACC Distribution, United Kingdom
+44 (0) 1394 389950
uksales@accpublishinggroup.com

The Scientific Committee of the issues List
Eve Blau (Harvard GSD), Maurizio Carta (University of Palermo), Eva Castro (Architectural Association London) Alberto Clementi (University of Chieti), Alberto Cecchetto (University of Venezia), Stefano De Martino (University of Innsbruck), Corrado Diamantini (University of Trento), Antonio De Rossi (University of Torino), Franco Farinelli (University of Bologna), Carlo Gasparrini (University of Napoli), Manuel Gausa (University of Genova), Giovanni Maciocco (University of Sassari/Alghero), Antonio Paris (University of Roma), Mosè Ricci (University of Trento), Roger Riewe (University of Graz), Pino Scaglione (University of Trento).

LISt Lab is an editorial workshop, based in Europe, that works on contemporary issues. LISt Lab not only publishes, but also researches, proposes, promotes, produces, creates networks.

LISt Lab is a green company committed to respect the environment. Paper, ink, glues and all processings come from short supply chains and aim at limiting pollution. The print run of books and magazines is based on consumption patterns, thus preventing waste of paper and surpluses. LISt Lab aims at the responsibility of the authors and markets, towards the knowledge of a new publishing culture based on resource management.